Nicolas Flamel

Nicolas Flamel

His Exposition of the Hieroglyphicall Figures (1624)

Edited by
Laurinda Dixon

Routledge
Taylor & Francis Group

First published in 1994 by Garland Publishing, Inc.

This edition first published in 2018 by Routledge
2 Park Square, Milton Park, Abingdon, Oxon, OX14 4RN
and by Routledge
52 Vanderbilt Avenue, New York, NY 10017, USA

Routledge is an imprint of the Taylor & Francis Group, an informa business

Publisher's Note
The publisher has gone to great lengths to ensure the quality of this reprint but points out that some imperfections in the original copies may be apparent.

Disclaimer
The publisher has made every effort to trace copyright holders and welcomes correspondence from those they have been unable to contact.
A Library of Congress record exists under ISBN:

ISBN 13: 978-0-367-18897-9 (hbk)
ISBN 13: 978-0-367-18900-6 (pbk)
ISBN 13: 978-0-429-19907-3 (ebk)

ENGLISH RENAISSANCE HERMETICISM
VOL. 2

NICOLAS FLAMEL

His Exposition of the Hieroglyphicall Figures (1624)

GARLAND REFERENCE LIBRARY
OF THE HUMANITIES
VOL. 979

ENGLISH RENAISSANCE HERMETICISM

STANTON J. LINDEN
Series Editor

THOMAS C. FAULKNER
Textual Consultant

JEAN D'ESPAGNET'S *ENCHYRIDION PHISICAE RESTITUTAE; OR, THE SUMMARY OF PHYSICKS RECOVERED* (1651), WITH THE *ARCANUM HERMETICAE PHILOSOPHIAE*
edited by Thomas S. Willard

BASIL VALENTINE HIS TRIUMPHANT CHARIOT OF ANTIMONY, WITH ANNOTATIONS OF THEODORE KIRKRINGIUS (1678)
edited by L.G. Kelly

THE MIRROR OF ALCHIMY
Composed by the Thrice-Famous and Learned Fryer, Roger Bachon (1597)
edited by Stanton J. Linden

ALCHEMICAL POETRY 1575–1700
From Previously Unpublished Manuscripts
edited by Robert M. Schuler

NICOLAS FLAMEL
His Exposition of the Hieroglyphicall Figures (1624)
edited by Laurinda Dixon

NICOLAS FLAMEL

His Exposition of the Hieroglyphicall Figures (1624)

edited by

Laurinda Dixon

GARLAND PUBLISHING, Inc.
New York & London / 1994

Library of Congress Cataloging-in-Publication Data

Flamel, Nicolas, d. 1418.
   [Figures hierogliphiques.   English]
   Nicolas Flamel : his exposition of the
hieroglyphicall figures (1624) / edited by Laurinda
Dixon.
      p.   cm. — (English Renaissance hermeticism ;
v. 2) (Garland reference library of the humanities ,
vol. 979)
   Includes bibliographical references and index.
   ISBN 0–8240–5838–0 (alk. paper)
   1. Alchemy—Early works to 1800.   I. Dixon,
Laurinda S.   II. Title.   III. Series.   IV. Series:
Garland reference library of the humanities , vol. 979.
QD25.F613   1994
540'.1'12—dc20   94–8101
                                                              CIP

Printed on acid-free, 250-year-life paper
Manufactured in the United States of America

CONTENTS

LIST OF ILLUSTRATIONS

GENERAL INTRODUCTION

The study of the nature, origin, dissemination, and influence of hermetic thought has been an important subject of Anglo-American and Continental scholarship for more than three decades. At present, despite the recent deaths of pioneers like Frances Yates and D. P. Walker, it continues to attract the attention of scholars internationally and in a variety of fields: literature, history, philosophy, religion, art history and iconography, and the history of science and medicine. Evidence of this interest is abundant and has taken the form of an increasing number of scholarly articles, monographs, and collections of essays; the organization of international conferences; and the publication of several specialized journals devoted to this subject. However, it is generally acknowledged that research in hermeticism suffers from an acute lack of reliable primary texts, a deficiency which this series is intended to alleviate with respect to the English Renaissance and seventeenth century.

English Renaissance Hermeticism is a series of new, authoritative editions of rare hermetic and alchemical texts which, with few exceptions, have not been reprinted since their original publication in England in the sixteenth and seventeenth centuries. It includes treatises written originally in English, as well as early translations of works by Continental authorities–past and contemporary–that were widely read and influenced English Renaissance thought and art in important ways. In addition to prose treatises, the series will include at least one volume of previously unpublished alchemical poetry.

Authors and titles to appear in *English Renaissance Hermeticism* have been selected because of their intrinsic importance and also to demonstrate the wide range of alchemical and hermetic thought that flourished in the Renaissance, even as the era of the "New Science" was dawning. For example, the physical transmutation of metals, explanations of the preparation of the philosopher's stone, iatrochemistry, Paracelsian expositions, varieties of spiritual and mystical alchemy, Rosicrucianism and Cabalism are represented. A partial listing of authors includes Basil Valentine, Jean d'Espagnet, Roger Bacon, Oswald Croll, Nicholas Flamel, Robert Fludd, Eirenaeus Philalethes and Hermes Trismegistus. Thus the series will include works that are among the least accessible and most important for interdisciplinary research, and are intended to comprise a core collection of texts which are vital as background to the study of Renaissance literature, intellectual history, science, and philosophy.

Primary place in this series will be given to treatises on *hermeticism* rather than *hermetism*. According to a scholar recently concerned with defining varieties of Renaissance occultism, the latter narrowly designates religious and philosophical writings attributed to Hermes Trismegistus and their interpretation throughout history. *Hermeticism,* more eclectic and broader in scope, refers to an "amorphous body of notions and attitudes deriving not merely from Hermes but also from the mystical side of Plato and his Neoplatonic successors and from such other esoteric systems as the numerology of Pythagoras and the Jewish cabala."[1] It is this syncretic body of knowledge, belief, and speculation that provides a basis for the theory and practice of magic, astrology, and, especially, alchemy with which most of these Renaissance writers are concerned.

Finally, volumes appearing in the *English Renaissance Hermeticism* series will have several features in common: all will be edited by scholars active in the field of hermetic studies; all will have new, reset texts carefully transcribed and verified against original editions (facsimile reproduction will be used only for title pages and illustrations). Texts will be edited conservatively, preserving old spelling (except for ordinary normalizations) and

---

[1]See Wayne Shumaker, "Literary Hermeticism: Some Test Cases," in *Hermeticism and the Renaissance: Intellectual History and the Occult in Early Modern Europe,* ed. Ingrid Merkel and Allen G. Debus (Washington: Folger Shakespeare Library, 1988), 293–94.

punctuation. All volumes will feature critical introductions and full scholarly notes, bibliographies, and indices, and will be uniformly bound.

Stanton J. Linden
General Editor

INTRODUCTION

Nicolas Flamel's *Exposition of the Hieroglyphicall Figures which he caused to bee painted upon an Arch in St. Innocents Church-yard in Paris*, first published in a French edition of 1612, was destined to inspire debate and conjecture not only in its own century, but for three hundred years thereafter. Indeed, the name Nicolas Flamel has attained nothing less than mythic proportions in the annals of the history of alchemy, and the nature of his contribution continues to be the subject of romantic speculation to this day. Was Flamel a successful medieval alchemist who, having attained the secret of transmutation, proceeded to endow half the city of Paris with charitable works, or was he a humble scribe whose best known work, the *Exposition of the Hieroglyphicall Figures*, was actually written by its publisher P. Arnauld de la Chevalerie under the pseudonym "Eirenaeus Orandus"?

Nicolas Flamel was often cited as a medieval authority in alchemical books printed after 1612, and his image as a legendary adept remained untarnished until the eighteenth century. His true historical identity was first questioned in 1761 by the Abbé Villain, who claimed that the tale of Flamel's charmed life could not be supported by historical evidence, and that the publisher P. Arnauld de la Chevalerie was probably both the author of the *Exposition of the Hieroglyphicall Figures* and the originator of the Flamel legend.[1] Villain's controversial stance elicited a flood of protest led by Antoine-Joseph Pernety, who vehemently defended the image

---

[1] Etienne F. Villain, *Histoire critique de Nicolas Flamel et de Pernelle sa femme, recueillie d'actes anciens qui justifient l'origine et la médiocrité du leur fortune contre les imputations des Alchimistes* (Paris, 1761) and idem, *Essai d'une histoire de la paroisse de Saint Jacques de la Boucherie* (Paris, 1758).

of Flamel as a medieval alchemist whose great wealth was attained by finding the philosopher's stone.[2] Though largely based upon claims made in the author's introduction to the *Exposition of the Hieroglyphicall Figures*, Pernety's thesis was defended (with some embellishments) throughout the nineteenth century, with Albert Poisson and Louis Figuier supporting the traditional view of Flamel as the true author of the *Exposition*, and Auguste Valet de Viriville, among others, taking the opposite stance.[3] Throughout the centuries the legend has been enlarged and embellished, sometimes to an absurd extent. As early as the seventeenth century, travelers claimed that Flamel and his wife Perrenelle were still living and working in India, both having attained the ripe old age of nearly four hundred years. The mystical Rosicrucians made Flamel into a veritable archetype, adding the quality of immortality to a legend already exaggerated beyond belief. During the eighteenth century, at the height of the argument between Villain, Pernety and their various adherents, deluded spectators claimed they saw Flamel, Perrenelle and their son attending a performance of the Paris Opera accompanied by an artist who was sketching their portraits.[4] Conjecture about Nicolas Flamel's true nature continues in the present day, particularly among French occultists, to whom he is a national hero.[5]

Most of the accounts of Flamel's life which depict him as an accomplished alchemist are based upon the Introduction to the *Exposition of the Hieroglyphicall Figures*, reputed to be Flamel's own

---

[2]See Antoine-Joseph Pernety, *Lettre rétablissant les faits contre les préjugés de l'abbé Villain*, in M. Freron, *L'Année Littéraire* 7 (1758), Letter XI; and idem, *Histoire critique de Nicolas Flamel* (Paris, 1761).

[3]See Albert Poisson, *Histoire de l'Alchimie; XIVe siècle: Nicolas Flamel, sa vie, ses fondations, ses œuvres; suivi de la réimpression du Livre des figures hiéroglyphiques et de la lettre de Dom Pernety à l'abbé Villain* (Paris, 1893); Louis Figuier, *L'Alchimie et les alchimistes* (Paris, 1860); Auguste Vallet de Viriville, "Quelques recherches sur Nicolas Flamel," in *Revue Française* 3 (1837), and idem, "Des ouvrages alchimiques attribués à Flamel," *Mémoires de la Société impériale des Antiquaires de France* 3 (1857).

[4]M. Caron and Serge Hutin, *The Alchemists* (New York and London, 1961), 19; also Arthur Edward Waite, *Alchemists Through the Ages* (1888; reprint, New York, 1970), 115–18.

[5]See, for example, Leo Larguier, *Le Faiseur d'or Nicolas Flamel* (Paris, 1936); Eric Muraise, *Le Livre de L'ange. Histoire et légende alchimique de Nicolas Flamel* (Paris, 1969); and Gilette Ziegler, *Nicolas Flamel, ou le secret du Grand Oeuvre* (Paris, 1971).

narrative. The tale begins humbly with the author as a Parisian notary living in a house near St. Jacques de la Boucherie, a church destroyed by revolutionary violence in 1791. In his professional capacity as a scribe, Flamel admits to having some knowledge of the "Bookes of the Philosophers," and to acquiring for very little money a "guilded Booke, very old and large . . . all engraven with letters, or strange figures." Though the majority of the writing in this book was in an obscure ancient language, the title page bore a decipherable inscription attributing it to "Abraham the Jew." According to his account, Flamel became obsessed with deciphering the enigmatic illustrations, showing them to his wife and friends and even painting them upon the walls of his house. Determined to unlock their secret, he undertook a pilgrimage to Spain, in hopes of finding a "Jewish Priest in a synagogue" who could help him. On the way back from Santiago da Campostella, Flamel met "Master Canches," a converted Jew, physician and alchemical adept, who deciphered the meaning of the book. The two set out together for France, but, sadly, Canches died of a mysterious illness en route. Luckily for Flamel, Canches had already revealed the secret of transmutation to his willing student, and, upon his return to Paris, Flamel and his faithful wife Perrenelle successfully transmuted base metals three times. With the riches gleaned from their successes, Nicolas and Perrenelle claimed to have endowed fourteen hospitals, built three new chapels, enriched and repaired seven churches in Paris and nearly as many in the city of Bologne and eased the burdens of countless widows and orphans. The text and illustrations of the *Exposition of the Hieroglyphicall Figures* claim to elucidate the design of a sculpted tympanum (described in the text as an "arch") given in 1413 to the Cemetery of the Innocents in the parish of St. Jacques de la Boucherie by Nicolas and Perrenelle Flamel. The tympanum, no longer extant, was supposedly inspired by the text and images of the *Book of Abraham the Jew.*

The actual events of Flamel's life, uncovered by the industrious Abbé Villain, are somewhat less glamorous than the legend. It appears that a public scribe by the name of Nicolas Flamel actually did live in Paris in the late fourteenth and early fifteenth centuries. The year 1330 is usually cited as his possible birth date, though the actual year is debatable, as is the place of birth, which is sometimes listed as Pontoise and sometimes Paris. The scribe Flamel ran two shops built against the wall of the church of Saint Jacques, served as a church warden in his parish and married Perrenelle, who brought with her the fortunes of two previous husbands, in 1368.[6] There is no doubt that the couple was wealthy, owning several properties and financially contributing to many churches and hostels. During Perrenelle's lifetime, she and her husband commissioned several sculpted works for churches, an activity that Nicolas continued after her death. Among these commissions was the tympanum for the Charnel House of the Innocents whose sculpted decorations are the subject of the illustrated *Exposition of the Hieroglyphicall Figures* and which, according to Langlet du Fresnoy, stood until the year 1742.[7] Scholars, in their attempts to burst the alchemical bubble, have sought to explain the couple's obvious wealth by suggesting that illegal business dealings with Jews supplemented their honest earnings.[8] There were, in fact, several expulsions of Jews from France during Flamel's lifetime—in 1346, 1354, 1380, 1382 and 1393—which would have presented an unscrupulous entrepreneur with the opportunity to claim, in the name of the King, lands and moneys left by the Jews.[9] There are, however, other possible explanations for the apparent flamboyant spending of Nicolas and Perrenelle. The couple was childless and could afford to disperse their wealth in charitable works and commemorative art. As was the case with similar wealthy childless couples throughout history, they would have been compelled to assure their continued veneration after death by supporting the Church during their lives.[10]

---

[6]See Caron and Hutin, 16.

[7]See Waite, 108–9.

[8]Poisson, *Histoire*, 94.

[9]See Claude Gagnon, *Description du "Livre des figures hieroglyphiques" attribué à Nicolas Flamel; Suivie d'une réimpression de l'édition originale et d'une reproduction des sept talismans du "Livre d'Abraham," auxquels on a joint le "Testament" authentique dudit Flamel* (Montreal, 1977), 58.

[10]A famous instance of a childless couple leaving their wealth to posterity in the form of a work of art is the case of Jodocus Vyd and Elisabeth Borluut, who commissioned the magnificent *Ghent Altarpiece*, painted by the Van Eycks, in partial fulfillment of their legacy. See Elisabeth Dhanens, *Van Eyck: The Ghent Altarpiece* (New York, 1973),

There is no doubt that Nicolas Flamel once lived, for his tombstone, originally placed in the cemetery of St. Jacques, now rests in the Musée de Cluny, having been saved from an ignominious existence as a cutting board in a Parisian grocery.[11] Flamel's will is dated 22 November 1416 and his death is believed to have taken place on 22 March of the following year. Though generous, Flamel's actual bequest does not suggest anything approaching the extraordinary wealth claimed in the *Exposition of the Hieroglyphicall Figures*.[12] Any fortune he may have acquired during his life probably came from his profession as a scribe (a prestigious and highly-paid occupation before the invention of printing) and the personal fortune of the twice-widowed Perrenelle rather than from the philosopher's stone. Indeed, Perrenelle left her husband 5,300 Tours pounds at her death in 1397, a sum that was hotly contested in the courts by her sister and brother-in-law.[13] Thus, historical circumstance has preserved many incontestable facts about Nicolas Flamel, making him one of the best-documented medieval figures in the history of alchemy. Nowhere, however, does the historical record mention alchemy, pharmacy, medicine, or the acquisition by Nicolas Flamel of any education beyond that required of a copyist and notary. No original manuscript version of the *Exposition of the Hierglyphicall Figures* or the *Book of Abraham the Jew* exists, and no existing alchemical treatises written before the late sixteenth century cite Nicolas Flamel as a medieval source. Flamel was a real person, and he may have dabbled in alchemy, but his reputation as an author and immortal adept must be accepted as an invention of the seventeenth century.

22–6.

[11]Musée des thermes et de l'hôtel de Cluny, *Catalogue général* (Paris, 1922), 1:105, no. 574.

[12]Hutin, 16.

[13]*Ibid.,* 17.

Flamel's Treatise in Historical Context

The seventeenth century was an era poised at the edge of modernity, yet also mired in medieval tradition. At the same time that Robert Boyle was formulating the theories that would sound the death knell to alchemy, there occurred a genuine resurgence of interest in early alchemical literature.[14] Hence, Flamel's *Exposition of the Hieroglyphicall Figures* as well as Artephius' *Secret Booke* and *The Epistle* of John Pontanus that appeared with it in the 1624 English edition, were calculated to appeal to a wide audience of readers. Early alchemical treatises had always been available in manuscript form, and would continue to be so despite the advent of the printing press. The earliest texts, notably the *Turba of the Philosophers* and Hermes Trismegistus' *Emerald Tablet*, were published in the fifteenth century, and by the early sixteenth century most of the medieval treatises had appeared in print.[15] The seventeenth century concerned itself with compiling and publishing these early texts in large collections of treatises. The most notable and extensive of these to appear before the first 1612 printing of Flamel's *Exposition of the Hieroglyphicall Figures* was the *Theatrum chemicum*, an alchemical compendium printed at Ursel in 1602 with consecutive volumes appearing in 1613, 1622, 1659 and 1661. Almost simultaneously in 1601, the Leiden publisher Nicolas Barnaud printed *De occulta philosophia*, featuring several early texts also found in the *Theatrum.*[16] As a result, the works of me-

---

[14]See Allen G. Debus, *Alchemy and Chemistry in the Seventeenth Century* (Los Angeles, 1966) and Lynn Thorndike, *History of Magic and Experimental Science* (New York, 1958), vol. 8.

[15]See Lynn Thorndike, "Alchemy During the First Half of the 16th Century," *Ambix* 2 (1938):26–37; R. Hirsch, "The Invention of Printing and the Diffusion of Alchemical and Chemical Knowledge," in *The Printed Word: Its Impact and Diffusion* (London, 1978), 115–41; and Laurinda Dixon, *Alchemical Imagery in Bosch's "Garden of Delights"* (Ann Arbor, 1981), Appendix B, for a list of alchemical books printed between 1460 and 1515.

[16]See Thorndike, 8:154–202 for other collections, and John Ferguson, *Bibliotheca Chemica* (Glasgow, 1906), 2:436–40 for the table of contents of the *Theatrum Chemicum.*

dieval authors such as George Ripley, Petrus Bonus, Roger Bacon, Thomas Norton and Raymond Lull were better known in the early years of the seventeenth century than they had ever been. The creation of another medieval adept in the person of Nicolas Flamel was not questioned by either scholars of alchemy or the interested reading public, from whom there issued a great demand for early alchemical treatises in printed form.

Flamel's *Exposition of the Hieroglyphicall Figures,* though not an actual medieval treatise, does, in fact, read like one. The sculpted cemetery tympanum which the text claims to elucidate and which most editions of the *Exposition* illustrate, has no blatant alchemical elements that could not also have been interpreted Biblically by any pious layperson of the seventeenth century. The text acknowledges this fact, claiming that the hieroglyphics may represent "two things, according to the capacity and understanding of them that behold them. . . . " Furthermore, the very structure of Flamel's treatise is divided into two major parts entitled "Of the Theologicall Interpretations, which may be given to these Hieroglyphickes" and "The interpretations Philosophicall, according to the Maistery of Hermes." The liturgical bias is echoed in the text, which expresses a fervent tone of medieval Catholic piety that must have been quaintly appealing to the revivalist spirit of the seventeenth century. The author, who claims to have received knowledge of the mysterious hieroglyphics during a pilgrimage to Santiago da Compostella, adheres to the traditional medieval identification of the philosophers' stone with Christ, though not without "first asking leave of the Catholicke, Apostolicke and Romane Church." The allegorical view of distillation as comparable to Christian death and resurrection also appears, and the virtues of secrecy and piety are continually stressed, though, by the seventeenth century, alchemy was openly practiced and its parameters included both the practical and spiritual applications of the art.[17] The virtue of piety, though still important, was often considered secondary to the learning and industry of the practitioner. The theological-alchemical duality inherent in the *Exposition* is, in

---

[17]For a description of the virtues required of medieval alchemists, see Albertus Magnus, *Libellus de alchimia,* trans. V. Heines (Berkeley, 1958), 12–14.

essence, typical of the broader nature of medieval alchemy, which, like its sister disciplines medicine and pharmacy, could claim both practical and philosophical applications.

In addition to the obvious Christian medievalism of the *Exposition of the Hieroglyphicall Figures*, the work also reveals some characteristics typical of later alchemical texts. For example, the author occasionally drops the Christ-*lapis* simile, referring instead to a "king" as the personification of the philosopher's stone. This is typical of many post-Reformation treatises, which tended to favor the symbol of a secular king as the personification of the *filius philosophorum*.[18] The author also frequently resorts to Classical myths to illustrate his points, calling upon the shades of Jason and Hercules as exemplars of the super-human strength and courage required to successfully achieve transmutation. Though not unknown in the fourteenth century, the myths of the ancient Greeks were more frequently cited in alchemical treatises during and after the Renaissance, when Classical sources were revived and interpreted, thereby becoming acceptable subjects for writers and artists.

The text of the *Exposition* cites numerous authorities, a practice common in early scientific treatises. However, only those who lived before the fourteenth century appear and, among those, Classical Greek and Arabic adepts predominate. Thus, the claim of an Eastern, "Jewish" source for the *Exposition* is backed by a careful choice of cited material which includes acknowledgments to Democritus, Orpheus, Diomedes, Hermes Trismegistus, Artephius, Rosinus, Morienus, Pythagorus, Calid, Avicenna, the *Turba Philosophorum* and the *Cabala*. An exception is a quotation attributed to Lambsprinck, a medieval German adept whose treatise *De lapide philosophica* was first printed in 1599.[19] Interest in Arabic and Cabalistic imagery was revived during the Renaissance, when philoso-

---

[18]The allegorical figure of the "king" as the embodiment of the *lapis* appears in several alchemical emblem books: Petrus Bonus, *Pretiosa margarita novella*, ed. J. Lacinius (Venice, 1546); Lambsprinck, *De lapide philosophica libellus* (Frankfurt, 1625); Michael Maier, *Atalanta fugiens* (Oppenheim, 1618) and Johann Daniel Mylius, *Philosophia reformata* (Frankfurt, 1622).

[19]In the *Triga chemica* (Leiden, 1599). The best-known version of the work is the 1625 Frankfurt edition, which is accompanied by copious emblematic illustrations.

phers such as Pico della Mirandola and Cornelius Agrippa von Nettesheim began to incorporate Eastern elements into their writings. Thus, fascination with Cabalistic imagery and Arabic alchemy was more characteristic of the seventeenth century than the fourteenth which, we have seen, was a time when "Jewish" sympathies went best unacknowledged. The *Cabala* would have been accessible to the author of the *Exposition of the Hieroglyphicall Figures* through Johann Reuchlin's *De arte cabbalistica* (Hagenau, 1517) and Pico della Mirandola's "Apologia tredecim Quaestionum" (included in the 1557 *Opera* published in Venice). The influence of Cabalistic imagery upon early seventeenth-century alchemy appears in illustrated treatises such as Heinrich Khunrath's *Amphitheatrum sapientiae aeternae* (1602), Steffan Michelspacher's *Cabala* of 1616, and Michael Maier's *Arcana arcanissima* of 1614. These treatises, like Flamel's *Exposition*, made the avowed claim of merging Christian theology with Arabic wisdom. Later in the century, combined Biblical-Cabalistic imagery would become a mainstay of Rosicrucian authors such as Thomas Vaughan.[20]

The Illustrations

Any consideration of Flamel's *Exposition of the Hieroglyphicall Figures* must, of necessity, include a discussion of its illustrations. The dual alchemical-Christian nature of the images fits comfortably within the Northern European artistic tradition, which accepted the intrinsic symbolic duality of objects depicted in art as a common visual convention. However, the great multiplication of available alchemical sources that occurred in the seventeenth century requires that any illustrated alchemical book from that time be interpreted in light of both alchemical and artistic precedent, a task that could easily consume the lifetimes of several scholars. Perhaps a good place to begin would be with the emblem book phenomenon that swept the Western world in the sixteenth and seventeenth centuries. In fact, Flamel's *Exposition of the Hieroglyphicall Figures*, by its very title, is an homage to the origin of

---

[20]See Alan Rudrum, ed., *The Works of Thomas Vaughan* (Oxford, 1984).

the emblem tradition in Western European intellectual circles.[21] The first printed book to make a concerted attempt to unravel a series of enigmatic illustrations was the *Hieroglyphica of Horapollo*, first published in 1505. As its title suggests, this book purported to be a translation of the pictorial language of the ancient Egyptians and consisted of two volumes and 189 chapters devoted to deciphering the hieroglyphics. Believed to be a Greek translation of an Egyptian work, the *Hieroglyphica of Horapollo* was one of the ancient texts discovered in the early fifteenth century by Florentine Neoplatonists, who revered it as a genuine work of antiquity. By the sixteenth century, the *Hieroglyphica* had inspired a veritable craze for emblem books, all organized in essentially the same way. The standard form included a series of allegorical pictures or "emblems," each accompanied by a fable in Latin verse, the "stanza," and a short "motto" which expressed the moral lesson to be learned from the picture. Among the many fashionable "emblemata" which circulated among the literate classes was Alciati's *Emblematum liber* (Augsburg, 1551), which appeared in 130 editions between 1532 and 1781. The first English emblem book was Geoffrey Whitney's *A Choice of Emblems* printed in Leiden in 1586. The fashion for emblems peaked in the seventeenth century, attracting a wide popular reading audience. In response to public demand, publishers produced several fine emblem books based upon alchemical allegory and laboratory procedure. Flamel's *Exposition of the Hieroglyphicall Figures* was one of the first to be printed in the seventeenth century, though the way had been paved by the appearance of several high-quality illustrated texts in the previous century.[22] These works, in concert with Flamel's *Exposition*, provided early models for the great alchemical emblem books that came later, such as Michael Maier's *Atalanta Fugiens* (Oppenheim, 1617, 1618) and Mylius' *Philosophica Reformata* (Frankfurt, 1622).

---

[21]For a history of the emblem tradition, see Mario Praz, *Studies in Seventeenth-Century Imagery* (Rome, 1964).

[22]Such as the *Rosarium philosophorum* (Frankfurt, 1550); Hieronymus Reusner, *Pandora* (Basel, 1588); and Giovanni Battista Nazari, *Della tramutatione metallica* (Brescia, 1569, 1572, 1599).

The *Exposition of the Hieroglyphicall Figures*, like most illustrated alchemical treatises, is especially precise about color indications in the text. Explicit descriptions of the exact hues intended for each image are numerous, and fine distinctions between such colors as "orange," "reddish yellow," "yellow," "yellowish," "golden red" and "citrine," for example, are common. The emphasis on color is typical of all early printed alchemical books, even though they were illustrated solely by black and white engraved line drawings. This apparent dichotomy is resolved by the fact that color indications in alchemical texts, both printed and handwritten, were quite unrelated to aesthetic considerations or artistic expression, but rather were intended to reflect actual laboratory procedure. In the absence of a proper thermometer with which to measure the heat of their ingredients, early practitioners relied heavily upon color indications in their work.[23] In the early days of printing before mass color reproduction was possible, the owner of a wood block or engraved illustrated book would have added colors according to the instructions in the text in the manner of a modern children's coloring book. In fact, many popular printed alchemical books in collections today betray the remains of color applied after purchase by some past owner. Thus, the designation of color, either by hand-application of paint or by indications in the text, supplied important didactic information to readers.

The colorful illustrations in seventeenth-century alchemical books contributed to their critical success among the lay public. Alchemy, which had enjoyed a rich tradition of pictorial symbolism for the better part of two millennia, was a fertile field for artists and emblem makers. The catalogue of alchemical images was vast, and had been assembled by generations of adepts seeking to hide their art from the eyes of the ignorant and uninitiated. By the seventeenth century, many alchemical images had been used indiscriminately for hundreds of years and their origins had become buried in obscurity. Regardless of the name attached

---

[23]A device for measuring the temperature of the air was described by Francis Bacon in his *Novum Organum* of 1620. An instrument strong enough to withstand temperatures of molten metals in the hundreds of degrees farenheit, however, was not possible in the seventeenth century. See Daniel J. Boorstin, *The Discoverers* (New York, 1983), 370.

to the treatise in which they appeared, certain allegorical figures could be expected to turn up with regularity. By the time the *Exposition of the Hieroglyphicall Figures* appeared, the alchemical "king," "hermaphrodite" and "green lion," all of which appear in Flamel's text, had been repeated so often that they had become generic. When illustrated, often the only changes that time wrought in such symbolic images were expertise in spatial conception and drawing, and idealization of costume and human form in response to prevailing fashion.

On the other hand, a passage of alchemical text could inspire a visual interpretation of startling originality, producing images that were often bizarre and obscure in nature. Like the legend of Flamel himself, alchemical illustrations contained a blend of fantasy and reality, a merging of factual laboratory procedures with bizarre allegorical motifs that made them attractive to artists and illustrators.[24] As a result, alchemical images are as varied as they are numerous, indicating that artists had free reign to combine stock illustrations with images conjured from their individual imaginations. In many instances, the results are without precedent, resulting in an independent pictorial language that transcends the text. In all cases, the illustrations cannot be understood without a knowledge of the allegories and procedures of alchemy. The large, fold-out illustration of the archway of the Cemetery of the Innocents that accompanies most editions of Flamel's *Exposition* was not, however, formulated by traditional means. It was not invented by an author or artist, nor was it part of the catalogue of stock alchemical images, but purported to be an accurate reflection of the actual appearance of an existing medieval sculpted monument. This approach was unique among illustrated seventeenth-century alchemical books, and no doubt added a convincing archaeological appeal to the assertion of authenticity put forth by the author.

---

[24]See Dixon (1981); and idem, "Bosch's 'St. Anthony Triptych': An Apothecary's Apotheosis," *Art Journal* 44 (1984): 119–32; J. van Lennep, *Art et Alchimie* (Brussels, 1966); and A. A. A. M. Brinkman, *Chemie in de Kunst* (Amsterdam, 1975) for discussions of the influence of alchemical imagery on artists.

Flamel's Introduction

Flamel's *Exposition* begins with an "Introduction" that describes the seven emblematic images in the *Book of Abraham* and recounts the adventures of Nicolas and Perrenelle as they attempt to decipher them. During the eighteenth-century debate over the facts of Flamel's life, the narrative was interpreted by some as a pure allegory. The symbolic nature of the events recounted in the Introduction was offered as proof that the *Exposition* was newly composed in the seventeenth century and was never intended to be anything but an alchemical allegory from start to finish. True, the account of Flamel's supposed pilgrimage to Spain for the purpose of finding a Jewish interpreter of the hieroglyphics corresponds indirectly with historical fact. The Church of St. Jacques in Flamel's home parish of St. Jacques de la Boucherie was the place where pilgrims assembled before journeying to Compostella via the Rue de St. Jacques, the southern route out of Paris toward Spain.[25] On the other hand, the idea of a devout pilgrimage could also serve as an allegory of the phases of the alchemical work or as a simile for the piety, courage and hard work required of adepts in their quest for the philosophers' stone. For example, the place called "Montjoy," where Flamel stopped on his way to Spain, can be interpreted symbolically, as the word also refers to a pile of stones made by travelers to mark places of sanctuary along a pilgrimage route.[26] "Montjoy" can also serve as an alchemical allegory, for "rocks" and "mountains" often symbolized the furnaces and flasks in which ingredients were warmed and nurtured,[27] and appear as such in numerous illustrated alchemical emblem books.[28] In this

---

[25]*Les Oeuvres de Nicolas Flamel*, ed. Elie-Charles Flamand (Paris, 1973), 10.

[26]Ibid., 9. "Mount of Joy" was also the name given Monte Mario, the hill outside Rome where pilgrims received their first view of the city.

[27]See Roger Bacon, "Speculum alchimicum," in *Pharmacopoeia Londinensis*, ed. Nicolas Culpeper (London, 1659), 633.

[28]See, for example, Heinrich Khunrath, *Ampitheatrum sapientiae aeternae* (Hanover, 1609); Steffan Michelspacher, *Cabala* (Augsburg, 1616); Michael Maier, *Atalanta fugiens*

context, the return voyage by sea undertaken by Flamel and Canches, during which the converted Jew was afflicted with a fatal illness, could also be read as an allegory of the "humid work." Alternatively, alchemical authors often compared the uncertainty and peril of a sea voyage with the explosive, poisonous dangers of the laboratory.[29] Indeed, the *Exposition* itself employs this simile at the end of chapter 5, where the difficulties of *putrefactio* are compared to the "troubles of the sea." In the same way, Canches' death could serve as an allegory of the dissolution of matter essential for the continuation of the work. Even the date that Flamel gives for his first successful transmutation suggests more when interpreted within the symbolic framework of alchemy. As scholars have noted, 17 January 1382, the date recorded for Flamel's successful transmutation of mercury into silver, was actually a Friday, not a Monday as claimed in the text.[30] Allegorically, Monday, the day associated with the Moon and its characteristic metal silver, has more symbolic resonance within the astrological sub-context of Flamel's transmutation than does Friday, which belongs to the planet Venus and the metal copper. Perhaps the date was not an error on the part of the author, but a purposeful homage to the planet associated with the particular chemical operation described in the text. Scholars have also noted that the very names "Nicolas" and "Canches" mean, in Greek, "Vanquisher of the stone" and "sulphur" (actually, "the dry one").[31] Furthermore, the *Exposition* resonates throughout with alchemical numerical symbolism. The numbers three (the alchemical trinities of sun, moon, Mercurius; body, soul, spirit; animal, vegetable, mineral, etc.), seven (metals and planets), and their product twenty-one occur repeatedly. For example, Flamel mentions that he ornamented three chapels and repaired seven churches with the proceeds of his experiments. There are seven hieroglyphics in Abraham's book, Canches died after an illness lasting seven days, and Flamel claims to have transmuted base metal three times. The text of Abraham's

(Oppenheim, 1618), and others.

[29]See Michael Maier, *Viatorium* (Oppenheim, 1618), Emblem 5, which illustrates Magellan in his ship as a symbol of the perils of alchemy.

[30]According to Eugène Cansiliet, *La tour Saint-Jacques* (Paris, 1956), 20.

[31]Flamand, 9.

book has twenty-one folios and Flamel claims to have worked on the problem of transmutation for twenty-one years. In addition, the organization of the text of the *Exposition* into two major sections, the theological and hermetic, symbolizes the basic alchemical precept of the duality of opposites.

The most interesting section of the "Introduction" describes the hieroglyphics as they supposedly appeared in the mysterious *Book of Abraham*. The actual images were not illustrated in the original French edition of 1612, nor in the 1624 English edition, which limited its illustrations to the sculpted arch (fig. 1). Not until Salmon's editions of the late seventeenth century did an engraver attempt to realize visually the vibrant verbal descriptions in the text, placing the seven "hieroglyphicks" in boxes forming a rectangle around the original image of the sculpted tympanum (fig. 2). Salmon's images deserve note, for they were probably the prototypes for various extant manuscripts claiming to be the original *Book of Abraham* and for the *Uraltes chymisches Werk*, an eighteenth-century attempt to present the public with the "authentic" *Book of Abraham*.[32] More importantly, the verbal descriptions of the hieroglyphics are more characteristically alchemical than the details of the medieval tympanum that claim the bulk of the author's attention. The hieroglyphical images served as models for later illustrated alchemical books and, in some cases, were derived from previously published sixteenth-century emblem collections.

Three of the hieroglyphics are mentioned as appearing on "the first seventh," "the second seventh," and "the last seventh" pages of the *Book of Abraham*, designations that emphasize the symbolic power of the number seven and its multiplication by three that exists throughout the text. The first hieroglyphic is described as "a *Virgin*, and *Serpents* swallowing her up," a strange image that seems to have no conventional prototype outside of alchemical lore. In fact, this hieroglyphic was eliminated from Salmon's editions, which substituted a caducean image of two serpents devouring each other (illustrated sixth from the left in Salmon's diagram, fig. 2) for the original virgin-dragon picture. The image of the vir-

---

[32]Abraham Eleazar, *Uraltes chymisches Werk* (Leipzig, 1760).

gin and the dragon is, however, an ancient one that appeared in early manuscripts.[33] It was illustrated nearly contemporaneously with Flamel's *Exposition* in Michael Maier's *Atalanta fugiens* (fig. 3). The motto accompanying Maier's graphic image says: "The Dragon kills the woman, and she kills it, and together they bathe in the blood." The source for the image probably comes from the *Turba Philosophorum*, which describes the dual death and unification of the dragon and woman as an allegory of the union, death and decomposition of opposite elements necessary for the birth of the *lapis*.[34]

The next hieroglyphic, appearing on the "second seventh" page of Abraham's book (it is the first, reading left to right, in Salmon's illustration, fig. 2), showed "a *Crosse* where a *Serpent* was crucified." According to Poisson and Grillot de Givry this image refers to the fixing of the volatile.[35] However, it also reinforces the Christ-*lapis* simile that the author applies to the central figure in the cemetery tympanum in chapter 6. In this context, the crucified serpent refers allegorically to the putrefaction ("crucifixion") of the alchemists' materials necessary before their resurrection into a perfected, immortal substance. According to C. G. Jung, the "crucified brazen serpent" had its origin in Gnosticism, where it served as a healing image.[36] The Christian identification of the serpent with Christ also occurs in the Old Testament, which states that "Moses lifted up the serpent in the wilderness, even so must the Son of Man be lifted up." Melchior Cibinensis used the simile in his alchemical paraphrase of the Mass, published in 1617, which strongly emphasized the *lapis*-Christ identity.[37] The second hieroglyphic, then, emphasizes the dual hermetic-theological bias of the entire *Exposition*, as the snake, representing the materials in

---

[33]See, for example, Pseudo Thomas Aquinas, *De alchimia*, Leiden, Bibliotheek der Rijksuniversiteit, Cod. Voss. Chym. F. 29, fol. 95a.

[34]"Turba philosophorum," in *Artis auriferae* (Basel, 1593), 1:58. See H. M. E. de Jong, *Michael Maier's "Atalanta Fugiens": Sources of an Alchemical Book of Emblems* (Leiden, 1969), 312.

[35]Poisson, *Histoire*, 125; Emile Angelo Grillot de Givry, *Le Musée des sorciers, mages et alchimistes* (Paris, 1929), 396.

[36]C. G. Jung, *Psychology and Alchemy* (New York, 1968), 144.

[37]Published in Michael Maier, *Symbola aureae mensae* as "Symbolum" (Frankfurt, 1617).

their base, unperfected state, stands for Christ incarnate undergoing the crucifixion that preceded his purification and resurrection (as the *lapis*).

The account of the third seventh page (the last at the far right in Salmon's illustration, fig. 2) describes an image of "*Desarts,* or *Wildernesses* in the middest whereof ran many faire fountaines, from whence there issued out a number of *Serpents,* which ran up and downe here and there." Both Poisson and Grillot de Givry see this image as a symbol of the "multiplication" of the stone discussed in chapter 19. This is certainly the proper reading, for in this image the opposite qualities of humid and dry, the "desarts" and "fountaines," are intended to suggest the proper environment for the serpents to spawn and increase.[38] The alchemical fountain as a nurturing force equivalent to the "fountain of life" is illustrated in the *Rosarium philosophorum* of 1550 (fig. 4).

Descriptions of the other four hieroglyphics, said to appear on the rectos and versos of pages four and five of Abraham's book, follow the account of the three that ornament the seventh, fourteenth and twenty-first leaves. The first of these (second from the left in Salmon's version, fig. 2) depicts "*Mercury* of the *Pagans,*" and "a great old man, who upon his head had an *houre-glasse* fastened, and in his hands a hooke (or sithe) like *Death,* with the which, in terrible and furious manner, hee would have cut off the feet of *Mercury.*" Grillot de Givry identified this image as a symbol of "mortification."[39] However, Poisson's suggestion that it refers to the purification of mercury by lead, personified as Saturn with an hourglass and scythe halting the messenger of the gods by cutting off his feet, is the more likely interpretation.[40] Alchemically, the "cutting of the foot" was a common simile for the fixing of mercury.[41] The allegory is ingeniously illustrated in Nazari's *Della tramutatione Metallica* of 1572 as the mutilated god Mercury lacking both feet and hands (fig. 5).

---

[38]Poisson, 125; Grillot de Givry, 396.

[39]Grillot de Givry, 396.

[40]Poisson, 125. See Raymond Klibansky, Erwin Panofsky, and Fritz Saxl, *Saturn and Melancholy* (New York, 1964) for examples of the iconography of Saturn in the intellectual context of the Renaissance.

[41]See Antoine-Joseph Pernety, *Dictionnaire mytho-hermétique* (Paris, 1758), 383.

The verso of page four of Abraham's book (third from the left in Salmon's diagram, fig. 2) is described as a hieroglyphic depicting a flower with a blue stem, white and red petals and golden leaves. It grows on top of a mountain, where it is "sore shaken with the *North wind*" and surrounded by "*Dragons* and *Griffons.*" Poisson and Grillot de Givry interpret this image respectively as an allegory of the "two sperms, sulfur and mercury," and as the sublimation of mercury.[42] It is, however, an image with many meanings in the context of alchemical iconography. The "golden flower," like the "sacred rose," symbol of the womb of the Virgin Mary, was considered to be the allegorical birthplace of the *filius philosophorum*.[43] The flower appeared in manuscripts colored in the three symbolic hues of alchemy: blue (putrefaction), white (albification) and red (the *lapis*). These same colors appear ubiquitously throughout the alchemical literature in connection with the three major changes in the work. Perhaps their earliest mention occurs in the *Turba of the Philosophers*, which describes the alchemical process as "a chicken with a red crest, white body, and black feet."[44] As a symbol of the *lapis*, the golden flower also appears repeatedly in the *Mutus liber*, an alchemical book whose hieroglyphic illustrations feature a man and woman working together in the laboratory in recollection of Flamel and his wife Perrenelle.[45] The precarious situation of the alchemical flower as it is buffeted by north winds and menaced by monsters, undoubtedly refers to the dangers of the Saturnine *nigredo* upon which the work is based. Reusner's *Pandora* illustrates this concept slightly differently than does Salmon, but with the same inherent meaning (fig. 6). Here, three flowers labeled *rubeum* (red), *album* (white) and *Flos sapientum* (flower of wisdom) grow from the extended claw of a dark dragon which bites its own tail in imitation of the cyclic process of distillation. Both Reusner's image and Flamel's description depict the *lapis* as a golden flower which grows from

---

[42]Poisson, 125; Grillot de Givry, 396.

[43]The alchemical flower appears in the "Ripley Scrowle," London, British Library MS Add. 5025.

[44]*La Tourbe des philosophes* (Paris, 1683), 34; see also Salmon Trismosin, *Aureum vellus* (Rorschach am Bodensee, 1598-9), 246.

[45]See Altus, *Mutus liber* (La Rochelle, 1677).

danger and putrefaction to its ultimate perfection in imitation of the cycle of nature.

The recto of page five (Salmon's fourth hieroglyphic from the left, fig. 2) in Abraham's book is described as illustrating a rose tree growing against a hollow oak in a garden. At the foot of the plant "boyled a fountaine of most *white water*, which ranne head-long downe into the depths, notwithstanding it first passed among the hands of infinite people, which digged in the Earth seeking for it; but because they were blinde, none of them knew it, except here and there one which considered the *weight*." Poisson saw the clear water as the source of the stone, while Grillot de Givry likened the entire scene to the revivification of sublimated mercury.[46] Like the flowers of alchemy, however, the "pleasant garden of the philosophers" was often employed as a symbol for the alchemical laboratory, and the hollow tree signified a furnace or flask.[47] The philosopher's garden, complete with furnaces and flasks, appears as the frontispiece of Hieronymus Braunschweig's manual of distillation published in 1500 (fig. 7). Braunschweig's image shows several practitioners contentedly distilling among the pleasures of nature, surrounded by pools, flowers, animals and beautiful women all provided by nature's abundance. By contrast, the "infinite people" in Flamel's description fail to recognize the fountain of white water flowing among them, even though it passes before their very eyes and "among the [their] hands." They are blind to the true nature of the *lapis*, and symbolize those who labor in vain to find it. Indeed, false alchemists were often compared to "blind men without eyes," and warned to "walk in the light and fall not into the ditch of perdition as blind men."[48] The "infinite people" who do not see what is before their very eyes call attention to the nature of the *prima materia* of alchemy. The common stuff of creation was considered both base and noble, both precious and of little worth, "cast out upon the dung hill as a vile thing, and hidden from the eyes of the ignorant men."[49] Alchemists described

[46]Poisson, 125; Grillot de Givry, 396.

[47]See Pernety (1758), 207.

[48]Raymond Lull, "Clavicula" in *Aurifontina chymica*, ed. J. F. Houpreght (London, 1680), 166–67.

[49]Roger Bacon, *Radix mundi*, in *Pharmacopoeia Londinensis*, ed. William Salmon (1541,

it in contradictory terms as "most dear and valuable, yet vile and the most vile."[50] The *prima materia* was ubiquitous, found always and everywhere, yet unrecognized in its myriad debased forms. Those who discern its great worth are the most sage adepts, who, in the words of Flamel, "here and there . . . considered the *weight*."

The seventh and last hieroglyphic is described as picturing the biblical Massacre of the Innocents (third from the right in Salmon's edition, fig. 2), "the bloud of which *Infants* was afterwards by other Souldiers gathered up, and put in a great vessell, wherein the *Sunne* and the *Moone* came to bathe themselves." The Massacre of the Innocents is the only one of the seven hieroglyphics from Abraham's book to be also illustrated as part of the tympanum (fig. 1), though it receives no commentary in the alchemical chapters that follow the introduction to the text. The massacre is, however, discussed in chapter 1, which is devoted to the theological symbolism of the tympanum images. This placement is understandable considering the inarguable Biblical origin of the scene and its gruesome subject, which made it especially appropriate for a charnel house setting. Furthermore, historical circumstances may well have suggested such a subject for the sculpted decoration of a cemetery dedicated to "innocents" (children). In 1183, the Jews were expelled from France for supposedly having practiced Cabalistic blood rituals on a little boy named Richard in Pontoise, the supposed birthplace of Flamel. Hence, the image of pagans murdering innocent children could have been historically and politically significant both to the French and to the character of Flamel himself.[51]

Alchemically, the violent murder of children described in the seventh hieroglyphic was interpreted by Poisson as a reference to the extraction of two natures from the *prima materia* and by Grillot de Givry as a symbol of the preparation of silver.[52] However, the biblical subject clearly has another meaning in alchemical con-

---

London, 1696), 215.

[50]Hermes Trismegistus, "Tractatus aureus," in *Pharmacopoeia Londinensis*, 215.

[51]Gagnon, 58.

[52]Lennep, 179.

text. It symbolizes the stage of *putrefactio* or *calcination*, wherein the ingredients were heated to the point of "death," or oxidation, in their vessel. Various similes were commonly applied to this stage—mortification, hell, Saturn, *nigredo*, melancholia, leprosy, the end of the world—all of them signifying the violence and death connected with *putrefactio*. The associated image of the sun and moon bathing themselves refers to the processes of "solution" and "ablution," during which the substance in the flask was dissolved, cleansed and eventually whitened. This essential stage of the work was commonly illustrated in early manuscripts as a man and woman—the alchemical opposites in the guise of the "sun and moon," "king and queen," or "Adam and Eve"—embracing in long-necked "marriage flasks" or, more inventively, sitting nude in ponds and bathtubs.[53] Sometimes these images allude directly to the "chemical marriage," depicting the alchemical couple actually copulating in their cleansing bath waters.[54] The description in the *Exposition* recalls an earlier illustration from the *Rosarium philosophorum* (1550) which shows the king and queen undergoing the process of conjunction in a six-sided baptismal font (fig. 8). Related to the image in the *Rosarium*, and even closer to the description of the hieroglyphic in the *Book of Abraham*, is an illustration from Mylius' *Philosophia reformata*, which associates the king and queen with the sun and moon respectively (fig. 9). The infants' blood in which the sun and moon bathe was a common alchemical synonym for the *aqua permanens*. Its use in alchemy was based upon the symbolism of the cleansing blood of Christ in Christian liturgy.[55] Nonetheless, the author of the *Exposition* cautiously assures us that by "blood," he means not the blood of infants, but the "mineral spirit in metals" inherent in the sun, moon and mercury.

---

[53]See, for example, London, British Library, MS Sloane 2560, fol. 6 and *Ripley Scrowle*, MS Add. 5025; Biblioteca Apostolica Vaticana, MS Pal. Lat. 412, fol. 57; Leiden, Bibliotheek der Rijksuniversiteit, Cod. Voss. Chym. Q. 8. fol. 1; Munich, Deutsches Museum, MS no. 38150, n.p..

[54]See, for example, the *Pretiosissimum Donum Dei Georgium Anrach*, Paris, Bibliothèque de l'Arsenal, MS 975, fols. 13 & 14; and a sixteenth-century manuscript of the *Rosarium philosophorum*, St. Gallen, Stadtbibliothek Vadiana, MS 394a, fols. 34 & 64, illustrated in S. K. de Rola, *Alchemy: The Secret Art* (New York, 1973).

[55]See C. G. Jung, *Mysterium Conjunctionis* (New York, 1963), 293.

The Chapters

The text of chapter 1 is devoted entirely to a traditional theological interpretation of the tympanum in the Cemetery of the parish of St. Jacques de la Boucherie. In twenty-five pages, this chapter succinctly states the aim of the book, which is to draw direct parallels between alchemical iconography and conventional Christian imagery, thus presenting alchemy as an exalted process bestowed and directed by God. The description of the now-destroyed tympanum in chapter 1 suggests no alchemical symbolism, and there is no part of it that could not have been interpreted by any devout fifteenth-century layperson as an expression of the familiar Christian subject matter that adorned the facades of most medieval churches and public buildings. Thus, the presence of the donors Nicolas and Perrenelle, presented to Christ and the angels by their patron saints, is a visual formula repeated in countless Renaissance paintings. Likewise, the details of the scene serve the purpose of Christian theology. The dragons represent sin and its consequences, and the resurrection of the dead, a subject especially well-chosen for a cemetery location, brings to mind the reward of the blessed on Judgment Day. As the author perfunctorily states, "the most ignorant may well know how to give it this interpretation."

The simplistic theological reading stated in the first chapter yields to a much more lengthy and complex alchemical allegorical interpretation taking up the next eight chapters and all of eighty-six pages. Chapter 2, entitled "The interpretations Philosophicall, according to the Maistery of Hermes," begins by challenging several points of conventional wisdom that the author claims are violated in the appearance of the tympanum. Though the visual anomalies mentioned here are not disturbing to the modern reader, they would have sounded warning bells in the minds of medieval viewers familiar with the immutable doctrines of liturgical iconography. The reader is asked, for example, to note that the figure of Saint Paul is on the right of Christ instead of on the left, and to question the fact that the donors at the feet of the two saints are clothed and not naked as they should be on the Day of Judgment. Chapter 2, like all the ensuing chapters, is devoted to alchemical matters and illustrated at the beginning with an enlarged detail of the tympanum which is then explained in the text. Here, the image under discussion shows a hand holding a pen and inkwell set within a stylized gothic architectural framework. The author explains that, though the uninitiated see the surface meaning as a reference to the donor Nicolas Flamel's occupation as a scribe, the initiated will recognize the image as a disguised representation of an alchemical furnace. This furnace, called the "habitation of the poulet," houses a flask containing the "*Philosophicall Egge*," which is gently warmed on a nest of ashes. The remainder of the chapter considers the function of the alchemical furnace in generating perpetual temperate heat. Like the natural warmth that creates life inside an actual hen's egg, the "temperate heat of their bath," must be gentle enough to vivify substances without burning them.

Chapter 3 is illustrated with an enlarged detail of two dragons, one of them winged, which the text interprets alchemically as the fixed and volatile principals. Villain, in an attempt to de-mystify the image, saw the two beasts as an eagle and a bull, traditional symbols for the evangelists John and Luke. He linked the two figures with the image of a man menaced by a lion which heads chapter 9. Villain saw the man and lion, then, as the conventional symbols of the other two evangelists Matthew and Mark, thereby obviating any alchemical inferences.[56] However, this interpretation does not correspond with the undeniable fact that the beast labeled by Villain as a bull, though hooved and eared, also has the wings that should belong to the "eagle." Furthermore, the "eagle" is not only wingless, but also beakless and clawless, and sports distinctly un-birdlike ears and toes. It is more likely that the two beasts are intended to be dragons or griffins, and are specifically described in the text as representations of the duality of opposites inherent throughout the alchemical work. The opposites are similarly represented as two fire-breathing dragons in the *Rosarium Philosophorum* (fig. 4), but they could appear as any pair that are attracted to each other in love, yet whose union is fraught with enmity and difficulty—cold and moist, sun and moon, black and white, sulphur and mercury, East and West or, more commonly,

---

[56]Villain, 108–9.

male and female.[57] This pair merges and produces the philosopher's stone, the *quintessence*, which then takes on a more noble form than either of its parents. The desired fusion of opposites is indicated in the *Exposition* by the appearance of the dragons themselves, who are shown as intertwined. The head of one is connected to the tail of the other in the manner of a circular *ouroborus*, the ancient symbol of circulation depicted as a serpent both devouring and giving birth to itself (fig. 6).

Classical references in the text relate the dragons of chapter 3 to the serpents that the infant Hercules strangled in his cradle and to the violent dragons that guarded the golden apples in the gardens of the Hesperides. In like manner, the text compares the difficulties of *putrefactio* to Jason's search for the golden fleece, Theseus' victory over Crete, and Cadmus' killing of the serpent of Mars. The equation of *putrefactio* with a battle between man and dragon is realized visually in Emblem 2 from Lambsprinck's *De Lapide Philosophica* (fig. 10).[58] During this stage, the dragons must be killed and blackened in an operation lasting forty days and accompanied by an "exceedingly great" stench. The importance of colors in both the scientific and visual realization of the alchemical process is stated emphatically, for "only he judgeth it to be such by the sight, and the changing of colours." The specific colors mentioned in the chapter heading and throughout the text are "yellowish, blew and black." Black is the symbol of *putrefactio*, of earth, rotting and death. "Bluish and yellowish" hues, however, signify that *putrefactio* is not yet completed and that the materials are not totally blackened. The appearance of orange or "half red" is disastrous, indicating that the ingredients have been burned beyond recovery. The text designates the colors of the dragons as yellowish and blue, signifying *putrefactio* in process, and the background (the "field") black, suggesting its successful completion.

Chapter 4 is headed by an image of a man and woman, Nicolas and Perrenelle, clothed in orange garments and placed against

---

[57]See Jung, *Mysterium Conjunctionis* for other pairings.
[58]See also the fifteenth-century *Aurora consurgens*, Zurich, Zentralbibliothek, Cod. Rhen. 172, fol. 20 and emblem 15 of Michael Maier, *Atalanta fugiens* (Oppenheim, 1618).

an azure and blue background. A continuous scroll bearing words referring to the terrible Judgment Day joins and encircles them. Villain, seeking to interpret the tympanum within a strictly liturgical context, saw the figures as two prophets rather than as a man and woman, and explained their function as foretellers of the judgment to come.[59] The alchemical interpretation put forth in the text explains that the resemblance of the man and woman to Nicolas and Perrenelle is only incidental, and that "our particular resemblance was not necessarily required, but it pleased the painter to put us there." More important is the definition of the two as the alchemical opposites of man and woman who, in marrying, signify the joining of the four elements necessary to produce the philosopher's stone. When united, they become the alchemical *hermaphrodite*, whose contrary sexual natures are converted into a unified body (fig. 11). The stage of whitening, or *ablution* is the main topic of the chapter. The alchemist's efforts to cleanse the putrefied ingredients are compared to Apollo vanquishing the python and to Theseus sowing the dragon's teeth in earth. The violent and dangerous *putrefactio* from which *albification* springs is suggested in the words of the sinuous scroll which encircles the pair and warns of the perilous Judgment Day. The colors mentioned throughout this chapter are significant. The "azure and blew" of the image background signifies that the *putrefactio*, having been fully achieved in blackness, is lessening. As in chapter 3, the orange garments of the figures indicate that the "digestion" of the bodies—their solution into a unified, undifferentiated substance—is not yet complete. White is the desired color that signifies successful cleansing of the ingredients, the resurrection of the "king" from death and blackness into life and redemption.

A second image of *albification* heads chapter 5. Here, a man dressed in orange, black and white clothing kneels before St. Paul, who holds a sword and wears a yellowish-white robe bordered with gold. The words in the scroll held by the man, identified as Nicolas Flamel by the letter "N" in the gothic tracery above him, pleads for forgiveness. The text explains that the sins of the penitent worshiper signify the blackness of *putrefactio* which is vanquished

---

[59]Villain, 108.

by *albification*. St. Paul's sword symbolizes the death and violence that accompanies *putrefactio*. The weapon will decapitate the penitent sinner just as the black "head of the crow" is cut off during the process of cleansing. Similar allusions to corporeal punishment and execution as similes for *putrefactio* are common in early alchemical literature. Decapitation appears, for example, as a symbol for *putrefactio* in the fifteenth-century *Book of the Holy Trinity*.[60] A similar image occurs in a manuscript copy of the *Aurora consurgens*, where a snake-tailed figure decapitates the unified material personified as the sun-man and moon-woman.[61] Chapter 5 does not dwell entirely upon the unsavory *putrefactio* stage, but continues to describe the process of "multiplication," achieved by repeated dissolving and coagulating. The five black "wreaths" that encircle St. Paul's sword indicate that the processes must each be repeated five times a month for five months. As in the preceding chapters, color symbolism plays an important part in the iconographical structure of the text of chapter 5. The kneeling man's orange, black and white clothing represents *putrefactio* and *albification*, and St. Paul's light yellow robe symbolizes the cleansing "virgin's milk." The appearance of diverse colors—citrine, green, red, yellow, blue and orange, called the "peacock's tail" in chapter 7—suggests the end of *putrefactio* and the beginning of *albification*.[62] It is the *"perfect redde* of the *vermillion,"* however, that signifies the desired end of the process and the attaining of the philosopher's stone.

---

[60]St. Gallen, Kantonsbibliothek Vadiana, MS 428, fol. 5 v.

[61]Pseudo Thomas Aquinas, *Aurora consurgens*, Zurich, Zentralbibliothek, Cod. Rhen. 172, fol. 27 verso. The *Aurora* also exists in manuscript copies in Paris, Bibliothèque Nationale, MS lat. 14006 and Leiden, Bibliotheek der Rijksuniversiteit, Cod. Voss. Chym. 29. It was published in J. Rhenanus, *Harmoniae inperscrutabilis chymico-philosophicæ* (Frankfurt, 1625) and in *Artis auriferae* (Basel, 1593), 1:185–246.

[62]The *cauda pavonis* is described in many alchemical texts. See G. Ripley, *A Treatise on Mercury*, in *Aurifontina chymica*, ed. J. F. Houpreght (London, 1680) and Roger Bacon, *The Philosopher's Stone* (London, 1739). The "peacock's tail" is illustrated in several manuscript copies of S. Trismosin's *Aureum vellus*: Berlin, Staatliche Museum, Kupferstichkabinett MS 78D3; London, British Museum, Harley MS 3469; Oxford, Bodleian Library, Bod. MS Ash. 1395; Nuremberg, Germanisches Nationalmuseum, MS 1465b; Paris, Bibliothèque Nationale, MS Français 12.297. The *Aureum vellus* was published in 1598 (Rorschach am Bodensee), 1604 (Basel) and 1613 (Paris).

The immortal tripartite nature of the stone is the topic of chapter 6, which is illustrated with an image of three "resuscitants" coming to life on Judgment Day surmounted by Christ the Judge and music-making angels. Here, the philosopher's stone, the *filius philosophorum*, is unquestionably synonymous with Christ. This equation is among the more common alchemical metaphors embodied in visual emblems, and has a long history traceable at least to the thirteenth-century *Codicillus* of Raymond Lull. The *Rosarium philosophorum* borrowed from this tradition in its representation of the victorious *lapis* as the risen Christ stepping from his sarcophagus (fig. 12).[63] Like the Holy Trinity of Father, Son and Holy Spirit, the alchemical trinities of body, soul and spirit—sun, moon and mercury—were "triple in name but one in essence."[64] The concurrent link of the alchemical process with redemption is also an ancient simile.[65] Martin Luther used the metaphor to justify the virtue of alchemy in religious devotion "by reason of the noble and beautiful likeness which it hath with the resurrection of the Dead on the Day of Judgement."[66] The two musical angels at Christ's feet are also meaningful in alchemical context. The iconographical significance of their instruments, the lute representing the realm of the soul and the heavens, and the bagpipe the earthly and sensual world, recall the fusion of body and soul inherent in both the alchemical process and the Last Judgement.[67]

Chapter 7 tells how "rubification," or reddening, must be obtained through feeding the product the "Virgin's milk." This substance, which Ruland defines as the "Mercurial Water of the Sages, after it has been purified from the unclean and Arsenical Sulphurs," is pictured graphically in fig. 83 of Stolcius de Stolcen-

---

[63]See C. G. Jung, "The Lapis-Christ Parallel" in *Psychology and Alchemy* (New York, 1968), 345–431.

[64]London, Wellcome Institute for the History of Medicine Library, Wellcome MS 2456, fol. 332.

[65]See H. S. Sheppard, "The Redemption Theme and Hellenistic Alchemy," *Ambix* 7 (1959): 42–6.

[66]See J. W. Montgomery, "Cross, Constellation, and Crucible, Lutheran Astrology and Alchemy in the Age of the Reformation," *Ambix* 11 (June 1963): 65–86.

[67]See Emanuel Winternitz, *Musical Instruments and Their Symbolism in Western Art* (London, 1967).

berg's *Viridarium chymicum* (fig. 13).[68] The detail of the tympanum at the head of the chapter pictures two angels, colored orange, on a violet and blue field. Again, the angels represent the alchemical opposites sulphur and mercury, fixed and volatile. Their orange color is not the orange of putrefaction mentioned in earlier chapters, however, but an iridescent "golden Citrine red," which indicates that digestion has been accomplished. This chapter, more than any preceding one, stresses colors as indications of the laboratory process and as symbols of the stages of the alchemical work. The progression from black to blue to violet, a color made by the addition of red to blue, signifies progressive cleansing and perfection culminating in red, the color of the transmuting agent. The "Peacock's taile," a veritable rainbow of colors, precedes the final reddening, the "*true red purple*," "*Poppey* of the *Hermitage*," and "*vermillion.*"[69]

Chapters 8 and 9 describe the culmination of the alchemical process. Chapter 8 is headed by an illustration of the right half of the tympanum in which St. Peter, clothed in *Citrine red*, presents the female donatrix who wears an orange dress and kneels at his feet. The woman is Perrenelle, who also symbolizes the philosopher's stone. The text explains that, at this stage of the process, the stone requires those things that women also desire, "*multiplication*" (i.e., children) and "a more rich *Accoustrement*" (i.e., better clothes). Specifically, the female wishes not only children, but also a new dress colored the same red as St. Peter's robe. At this time, the stone multiplies in quantity, quality and virtue by means of repeated dissolving and fixing until it attains the desired red color. The only risk is the loss of the stone by too intense heat. The perfect red finally appears in chapter 9, illustrated by a man "*red purple*," holding the foot of a winged red Lion who

---

[68]Martinus Rulandus, *A Lexicon of Alchemy*, trans. Arthur Edward Waite (1612; York Beach, Maine, 1984).

[69]Red is universally mentioned as the desired color of the philosopher's stone. See, for example, Thomas Norton, *Ordinall of Alchemy*, in *Theatrum chemicum Britannicum*, ed. Elias Ashmole (London, 1562), 1:56; Hermes Trismegistus, *Tractatus aureus* in *Pharmacopoeia Londinensis*, 222; Mylius, *Anatomiae auri sive tyrocinium medico-chymicum* (Frankfurt, 1628), fig. 3; Hieronymus Reusner, *Pandora* (Basel, 1588); and George Ripley, *Bosome Book of Alchemy*, in *Collectanea chymica*, ed. W. Cooper (London, 1684), 116.

"*seemes would ravish and carry away the man.*" Here, the fully perfect stone is compared to the red flying lion, incapable of destruction by either the powers of heaven or the zodiac. Here, also, the woman discussed in the previous chapter has cast off her old orange dress and taken on the "pure & cleere *skarlet.*" This red is described jubilantly as "most fair & all-perfect" and "sparkling and flaming." The acquisition of the "supercoelestiall" stone is accompanied by a final prayer and promise to use this great miracle for the benefit of the Christian faith and the glory of the realm.

The Text, Its Illustrations and Editions

Despite claims to the contrary, no trace of either the text or the illustrations of Flamel's *Exposition of the Hieroglyphicall Figures* exists that dates before the seventeenth century. The first appearance of the work was the Paris edition of 1612 printed by Arnauld de la Chevalier, of which the English edition of 1624, reprinted here, is a literal translation. As *Le livre des figures Hierogliphiques de Nicolas Flamel*,[70] it was first published with two other tracts, *Le Secret Livre* of Artephius (which Eirenaeus Orandus kept in the 1624 edition) and *Le Vray Livre de la Pierre Philosophale du docte Synesius* (replaced in the English collection by *The Epistle* of John Pontanus).[71] Scholars have tried in vain to identify an earlier manuscript, preferably a Latin one, that could have served as a source for the text and illustrations of the *Exposition.*[72] Indeed, there exist in Parisian libraries several handwritten, illuminated manuscript versions of both the *Exposition* and the so-called *Book of Abraham Eleazar (the Jew).* None, however, can be proven to date from before the seventeenth century.[73]

---

[70]The full title is *Le livre des figures Hieroglyphiques de Nicolas Flamel, ai/n/si, quelles sont en la quatriesme arche du Cymitiere des Innocens a Paris, avec l'explication d'icelles per le dit Flamel, traictant de la transmutation metallique, non jamais imprime.*

[71]An earlier mention of Flamel as an alchemist occurs in "Dionysius Zacharius, Opusculum philosophiae naturalis metallorum, cum annotationibus Nicolai Flamel" in *Theatrum chemicum* (Ursel, 1602), 804–901.

[72]Lennep, 177, notes that both Canseliet and Poisson believed in the existence of a manuscript source for the *Exposition.*

[73]Manuscripts of Flamel's *Exposition of the Hieroglyphicall Figures* and hand-illuminated

So powerful were the alchemical descriptions of the hieroglyphics in Abraham's book that they were reincarnated in the eighteenth century as the *Uraltes Chymishes Buch* which claimed to be the lost tome of Abraham Eleazar (the Jew).[74]

Ordinarily, the existence of a manuscript version that post-dates a printed text would be an unusual phenomenon. However, the making of alchemical manuscripts did not stop with the invention of the printing press, but continued unabated through the eighteenth century. Perhaps the inadequacy of engraved line illustrations to convey qualities of color and atmosphere was a contributing factor in alchemy's protracted manuscript tradition. Nonetheless, details of costume and artistic convention indicate that extant manuscripts claiming to be the original versions of Flamel's treatise and the *Book of Abraham* were perhaps copied from a seventeenth-century printed book (probably Salmon's *Bibliothèque des philosophes*) rather than vice-versa.

Despite its spurious origin, Flamel's *Exposition* enjoyed a popular afterlife in the alchemical book-publishing trade. After the first French edition of 1612 and a second in the same year, came the English translation of 1624 that is reprinted here. Subsequent French editions based upon the 1612 text appeared in 1659, 1660 and 1682, and a German translation was published in 1669 and reprinted in 1673 and 1751. In addition, Flamel's treatise was included in William Salmon's *Bibliothèque des philosophes chimiques* dated 1672–78, and was re-edited without illustrations by de Richebourg in 1740–54.[75] Another English translation of Flamel's *Exposition* appeared in Salmon's *Medicina practica* of 1691, 1692 and 1707.[76] The text of this edition claims to be "newly translated into English," and differs considerably from Orandus' 1624

---

versions of the *Book of Abraham the Jew* are housed in Paris, Bibliothèque Nationale, MS Français 19075 and MS Français 14765; and in Paris, Bibliothèque de l'Arsenal MSS nos. 3047 and 973. C. G. Jung claimed to have in his possession another manuscript entitled *Figurarum Aegyptiorum secretarum* whose pictures were "identical with those in Paris, Bibliothèque de l'Arsenal, MS no. 973." See Jung, *Psychology and Alchemy*, 276, note 101, figs. 23, 148, 157, 164, and idem, *Mysterium Conjunctionis*, figs. 4–7.

[74]See Ferguson, 1:2-3.

[75]The Salmon edition includes Flamel's treatise in vol. 1; the Richebourg edition places it in vol. 2, 195–262.

[76]The 1707 edition lacks illustrations.

version. The illustrations in Salmon's editions are notable because they include variations on the basic themes presented in the earliest French and English editions. Those that accompany the *Medicina practica*, though closely related to the original illustrations, are presented as several fully-realized narrative scenes occurring within separate and distinct landscape backgrounds (fig. 14). The illustrations from Salmon's *Bibliothèque des philosophes*, on the other hand, include realizations of the seven figures from the *Book of Abraham* placed around the periphery of the charnel house tympanum illustration from the first edition (fig. 2). The hermetic emblems from the *Book of Abraham* are merely described verbally in earlier editions, and their inclusion in Salmon's collection is worthy of consideration and comparison with the colorful textual portrayals in the English translation of 1624.

The first editions of Flamel's *Exposition* were so successful that other treatises entitled *Le Sommaire Philosophique, Le Desir Desiré, Le Grand Eclairissement de la Pierre Philosophale, Le Livre des Laveures, Le Breviaire* and *La Musique Chimique* soon appeared under the purported authorship of Nicolas Flamel.[77] As in the case of the *Exposition of the Hieroglyphicall Figures*, these works have been the subject of continual debate as to their authenticity. Most historians of science, however, believe them to be apocryphal, and follow the example of Lynn Thorndike, who eliminated Nicolas Flamel altogether from the definitive *History of Magic and Experimental Science*.

Publishing of Flamel's *Exposition* continued through the modern era. At the end of the nineteenth century, Albert Poisson re-issued a fourth edition of de Richebourg's 1740 version, and a modern reprint of this edition appeared in 1970 edited by René Alleau. In 1973, *Les oeuvres de Nicolas Flamel* appeared with a preface by Elie-Charles Flamand (Paris: Belfond) and in 1977 Claude Gagnon re-issued the original French edition of 1612 with scholarly commentary and illustrations of Bibliothèque de l'Arsenal manuscript no. 3047 (Montreal: L'Aurore). The most recent attempt to bring Flamel's *Exposition* to the reading public is a 1980

---

[77]Other treatises attributed to Flamel were published in Borel's *La Bibliotheca chimica* (1654); the *Musaeum hermeticum reformatum et amplificatum* (1677–78 and 1749); *Aurifontina chymica* (1680); and Manget, *Bibliotheca chemica curiosa* (1702).

reprint of an 1889 version of the first English edition, issued with a resuscitated preface by the Rosicrucian author W. W. Westcott and accompanied by illustrations from one of the seventeenth-century French manuscripts (Berkeley Heights, New Jersey).

Artephius' *Secret Book* and Pontanus' *Epistle*

The two treatises printed with the Flamel *Exposition of the Hiero-glyphicall Figures* in Orandus' 1624 English edition are Artephius' *Secret Booke* and John Pontanus' brief *Epistle*. Artephius has been identified as the Arabic poet and alchemist Al Toghrai, who was executed ca. 1119–20 or 1121–22, according to Ibn Khallikan. By virtue of the alchemical elixir, he supposedly lived a thousand and twenty-five years—even longer than the legendary Nicolas Flamel who, if one believes the legend that he still lives, would be only about 600 years old today.[78] Not only do the authors Artephius and Flamel share a similar legendary proclivity toward long life, but their respective treatises have several elements in common. Both works, for example, refer to Jason and the gardens of the Hesperides, both compare the alchemical work to a dangerous sea voyage and both dwell upon the powerful image of the alchemical opposites as a king and queen sharing a cleansing and dissolving bath. In addition, both texts repeatedly emphasize colors and color changes in the alchemical work. Artephius' *Liber secretus* (*Secret Book*) appeared with the Flamel treatise in the first French edition of 1612 and, after its translation in 1624, was printed at Amsterdam (1678) and Frankfurt (1685). The *Liber secretus* also appeared in Salmon's *Medicina practica* (1691) and Richebourg's *Bibliothèque des Philosophes Chimiques* (1740).

John Pontanus' *Epistle* is a brief testament praising Artephius. The author's name is believed to have been the pseudonym of Johann Brueckner, who was doctor of philosophy and medicine and professor of philosophy at Koenigsberg from 1544 to 1545. He served for a year as chair of medicine and physics in 1552, after which he moved to Jena and then to Gotha, where he was appointed physician to the Prince. Pontanus' final position was as personal physician to the Duke of Weimar, whom he accompanied to Vienna. On this journey Pontanus died on July 9, 1572, sus-

---

[78]Ferguson, 1:50–1.

pected of having been poisoned. He himself published nothing, and his works only appeared in print after his death. The Latin text of the *Epistle* is included in Sendivogius, *Lumen Chymicum Novum* (1624), in vol. 3 of the *Theatrum chemicum* (1659), and with a German translation in Johann Peter Gerhard's *Gedanken vom Feure* (Halle, 1750).[79]

In sum, both Artephius and Pontanus were excellent choices to complement the illustrated Flamel text. The *Secret Booke* of Artephius shares much of the allegorical imagery found in Flamel's *Exposition* and the Pontanus text, in lauding Artephius, also affirms the greatness of Flamel's *Exposition.*

Editorial Principles

Several illustrated editions of the Flamel, Artephius and Pontanus treatises were consulted in the preparation of this volume. The transcription was done from a microfilm of the copy in the Historical Library of the Harvey Cushing/ John Hay Whitney Medical Library of Yale University. The transcription was read against the British Library copy 1032.a.19. The commentary, however, elucidates only the first English edition of 1624. Explanatory notes referring to the 1612 French edition (New York Public Library copy KB1659) and to the second English translation of 1691 (copies held by the National Library of Medicine, Bethesda [WZ250S172 M1692] and the New York Academy of Medicine [RB1707]) appear only when a comparison of a word or phrase would seem to shed light on an element of the 1624 English edition.

The editorial slant of this edition is predominantly art historical, and annotations are concerned primarily with the images or descriptions of images that accompany the text, the visual tradition that produced them, and the impact of the Flamel "hieroglyphics" upon the alchemical emblem tradition. The same is true of the comparative illustrations included in the introduction. The biblical and Classical references in all three texts are considered in light of their narrative and symbolic impact upon the developing

---

[79]Ferguson, 2:212–13.

catalogue of sources for alchemical imagery. The pages that follow present these texts not as the personal confessions of a pious medieval adept or as a Rosicrucian testament, but as an evocation of the "golden age" of alchemy presented in both word and image within the structure of hermetic thought.

Acknowledgements

I am especially grateful to several helpful individuals who aided me in the preparation of this edition. I wish to thank *English Renaissance Hermeticism* series editor Stanton J. Linden for his aid and support in all phases of this project. He deserves special credit for recognizing the appropriateness of an art-historical context in the editing of the Flamel *Exposition*. My thanks also go to Lucille Linden for assisting with research in the British Library, to Rhonda L. Blair for identifying the Greek quotations, and to Thomas C. Faulkner and the Humanities Research Center at Washington State University for technical assistance, layout and typesetting. I am especially grateful to the staffs of the Historical Library of the Harvey Cushing/ John Hay Whitney Medical Library of Yale University, New York Public Library, New York Academy of Medicine, the National Library of Medicine, and the library at Syracuse University. To all my colleagues who worked with me on this project, I express my sincere appreciation and thanks.

Bibliography

Albertus Magnus. *Libellus de alchimia.* Trans. V. Heines. Berkeley, 1958.

Altus [Jacob Saulat]. *Mutus liber.* La Rochelle, 1677.

Aquinas, Thomas [pseudo]. *Aurora Consurgens: A Document Attributed to Thomas Aquinas on the Problem of Opposites in Alchemy.* Trans. R. F. C. Hull and A. S. B. Glover. Ed. Marie-Louise von Franz. New York, 1966.

Artis auriferae, auriferae artis, quam chemiam vocant antiquissimi authores. 2 vols. Basel, 1593.

Aurifontina chymica. Ed. J. F. Houpreght. London, 1680.

Berthelot, Marcellin. *Collection des anciens alchimistes grecs.* 3 vols. Paris, 1887–8.

Bibliothèque des philosophes chimiques. Ed. William Salmon, Paris, 1672–8.

Bonus, Petrus. *Pretiosa margarita novella.* Ed. Janus Lacinius. Venice, 1546; Ed. Arthur Edward Waite. London, 1894.

Boorstin, Daniel J. *The Discoverers.* New York, 1983.

Boschius, Jacobus. *Symbolographia, sive De arte symbolica sermones septem.* Augsburg, 1702.

Braunschweig, Hieronymus. *Liber de arte distillandi, de Simplicibus. Das buch der rechten kunst zu distilieren die eintzige ding.* Strasbourg, 1500.

Brinkman, A. A. A. M. *Chemie in de Kunst.* Amsterdam, 1975.

Campy, David de Planis. *L'Ouverture de l'escolle de philosophic transmutatoire metallique.* Paris, 1633.

Canseliet, Eugène. *La tour Saint-Jacques.* Paris, 1956.

Caron, M. and Serge Hutin. *The Alchemists.* New York and London, 1961.

Cartari, Vincenzo. *Le imagini de i dei de gli antichi.* . . . Lyons, 1581.

Cluny, Musée des Thermes et de l'Hôtel. *Catalogue général.* Vol. 1. Paris, 1922.

Collectanea chymica. Ed. W. Cooper. London, 1684.

Coudert, Allison. *Alchemy: The Philosopher's Stone.* Boulder, 1980.

Debus, Allen. *Alchemy and Chemistry in the Seventeenth Century.* Los Angeles, 1966.

Dhanens, Elisabeth. *Van Eyck: The Ghent Altarpiece.* New York, 1973.

Dixon, Laurinda S. *Alchemical Imagery in Bosch's Garden of Delights.* Ann Arbor, 1981.

_____. "Bosch's 'St. Anthony Triptych': An Apothecary's Apotheosis." *Art Journal* 44 (1984): 119–32.

Eleazar (Abraham the Jew). *Uraltes chymisches Werk.* Leipzig, 1760.

Ferguson, John. *Bibliotheca chemica: A Catalogue of the Alchemical, Chemical and Pharmaceutical Books in the Collection of*

1

the Late James Young of Kelly, and Durris. 2 vols. Glasgow, 1906.

Figuier, Louis. *L'Alchimie et les alchimistes.* Paris, 1860.

Flamel, Nicolas. *Le livre des figures hieroglyfiques de Nicolas Flamel, aisi quelles sont en la quatriesme arche du Cymetiere des innocens a Paris, avec l'explication d'icelles par le dit Flamel, traictant de la transmutation metallique, non jamais imprime. Traduit du latin en Français par P. Arnauld, seiur de la Chevalerie, gentle-homme poitevin.* Paris, 1612, 1659.

——————. *Le Livre de Nicolas Flamel contenant l'explication des Figures hiéroglyphiques qu'il a fait mettre aucimetier des SS. Innocens à Paris.* In *Bibliothèque des philosophes (chimiques), ou recueil des Oeuvres des Auteurs, les plus approuvez, qui ont écrit de la Pierre Philosophale . . . Avec un Discours, servant de prefaces sur la verité de la science . . . ,* ed. William Salmon, vol. 1. Paris, 1672.

——————. *Le livre des figures hieroglyphiques.* Paris, 1970.

——————. *Nicholai Flammel Hieroglyphica.* In *Medicina practica; or, the Practical physick . . . To which is added, the philosophick works of Hermes Trismegistus, Kalid Percisus, Geber Arabs, Artefius Longaevus, Nicholas Flammel, Roger Bacon, and George Ripley. All translated out of the best Latin editions . . . The whole compleated in three books. (With engravings).* Ed. William Salmon. London, 1692, 1707.

——————. *Les Oeuvres de Nicolas Flamel.* Ed. Elie-Charles Flamand. Paris, 1973.

——————. *Le Sommaire philosophique.* In *Trois anciens tractes en rytmes françaises.* Ed. J. Gohory. Paris, 1561.

Fludd, Robert. *Summum bonum.* . . . . Frankfurt, 1629.

Gagnon, C. *Description du "Livre des figures hiéroglyphiques" attribué à Nicolas Flamel. Suivie d'une réimpression de l'édition originale et d'une reproduction des sept talismans de "Livre d'Abraham," auxquels on a joint le "Testament" authentique dudit Flamel.* Montreal, 1977.

Geber (Jabir ibn Haiyan). *De alchimia.* Nuremberg, 1529.

Gesner, Conrad. *New Book of the Distillatyon of Waters.* London, 1565.

_____. *Quattre livres des secrets.* Paris, 1573.

Grillot de Givry, Emile Angelo. *Le Musée des sorciers, mages et alchimistes.* Paris, 1929.

Hirsch, R. "The Invention of Printing and the Diffusion of Alchemical and Chemical Knowledge." In *The Printed Word: Its Impact and Diffusion.* London, 1978, 115–41.

Hutin, Serge. *Histoire de l'alchimie: de la science archaique à la philosophie occulte.* Paris, 1971.

Jamsthaler, Herbrandt. *Viatorium spagyricum. Das ist: Ein gebenedeyter spagyrischer Wegweiser.* Frankfurt, 1625.

Jong, H. M. E. de. *Michael Maier's "Atalanta Fugiens": Sources of an Alchemical Book of Emblems.* Leiden, 1969.

Jung, C. G. *Mysterium Coniunctionis.* Trans. R. F. C. Hull. New York, 1963.

_____. *Psychology and Alchemy.* Trans. R. F. C. Hull. New York, 1968.

Kelly, L. G., ed. *Basil Valentine His Triumphant Chariot of Antimony.* New York, 1990.

Khunrath, Heinrich. *Ampitheatrum sapientiae aeternae solius ve-rae.* Hanau, 1609.

Klibansky, Raymond, Erwin Panofsky and Fritz Saxl. *Saturn and Melancholy.* New York, 1964.

Lambsprinck. *De lapide philosophico.* Leiden, 1599; Frankfurt, 1625.

Larguier, Leo. *Le Faiseur d'or Nicolas Flamel.* Paris, 1936.

Lennep, J. van. *Art et alchimie: Etude de l'iconographie hermétique et de ses influences.* Brussels, 1971.

Libavius, Andreas. *Alchymia.* Nuremberg, 1529; Frankfurt, 1606.

Linden, Stanton J., ed. *The Mirror of Alchimy, Composed by the thrice-famous and learned Fryer, Roger Bachon (1597).* New York, 1992.

Lull, Raymond. *Le Codicille.* Trans. L. Bouyssou. London, 1953.

Maier, Michael. *Arcana arcanissima hoc est Hieroglyphica Aegypt-io-Graeca.* [London ?], 1614.

_____. *Atalanta fugiens.* Oppenheim, 1618.

_____. *Jocus severus.* Frankfurt, 1617.

_____. *Septimana philosophica.* Frankfurt, 1620.

_____. *Symbola aureae mensae duodecim nationum.* Frankfurt, 1617.

_____. *Tripus aureus.* Frankfurt, 1618.

_____. *Viatorium.* Oppenheim, 1616.

Michelspacher, Steffan. *Cabala, Spiegel der Kunst und Natur.* Augsburg, 1616.

Montgomery, J. W. "Cross, Constellation, and Crucible, Lutheran Astrology and Alchemy in the Age of the Reformation." *Ambix,* 11 (1963): 65–86.

Muraise, Eric. *Le Livre de L'ange. Histoire et legende alchimique de Nicolas Flamel.* Paris, 1969.

Musaeum hermeticum. Frankfurt, 1625.

Mylius, Johann Daniel. *Anatomiae auri sive Tyrocinium medico-chymicum.* Frankfurt, 1628, 1662.

——————. *Opus medico-chymicum.* Frankfurt, 1618.

——————. *Philosophia reformata continens libros binos.* Frankfurt, 1622.

Nazari, Giovanni Battista. *Della tramutatione metallica sogni tre.* Brescia, 1599.

Obrist, Barbara. *Les Débuts de l'imagerie alchimique (XIVe–XVe siècles).* Paris, 1982.

Opera omnia chemica. Cassel, 1649.

Pernety, A. J. *Dictionnaire mytho-hermétique dans lequel on trouve les allégories fabuleuses des poètes, les métaphores, les énigmes et les termes barbares des philosophes hermétiques expliqués.* Paris, 1787.

——————. "Essai d'une histoire de la paroisse Saint-Jacques de la Boucherie." In M. Ferron, *L'Année Littéraire* 7 (1758).

——————. *Histoire critique de Nicolas Flamel.* Paris, 1761.

_____. "Lettre rétablissant les faits contre les préjugés de l'abbé Villain." In M. Ferron, *L'Année Littéraire* 7 (1758).

Pharmacopoeia Londinensis. Ed. Nicolas Culpeper. London, 1659.

Pharmacopoeia Londinensis. Ed. William Salmon. London, 1696.

Poisson, A. *Histoire de l'Alchimie, XIVme siècle: Nicolas Flamel, sa vie, ses fondations, ses œuvres; suivi de la réimpression du Livre des figures hiéroglyphiques et de la lettre de Dom Pernety à l'abbé Villain.* Paris, 1893.

Praz, Mario. *Studies in Seventeenth-Century Imagery.* Rome, 1964.

Read, John H. *Prelude to Chemistry.* London, 1936.

Reusner, Hieronymus. [Franciscus Epimetheus, pseud.]. *Pandora: Das ist, die edelste Gab Gottes, oder der Werde und heilsame Stein der Weysen.* Basel, 1588.

Ripley, George. *Opera omnia chemica.* Kassel, 1649.

Rola, Stanislas K. de. *Alchemy: The Secret Art.* New York, 1973.

_____. *The Golden Game: Alchemical Engravings of the Seventeenth Century.* New York, 1988.

Rosarium philosophorum. Secunda pars alchimiae. In *De alchimia opuscula complura.* Basel, 1550.

Rulandus, Martinus. *A Lexicon of Alchemy* (1612). Trans. Arthur Edward Waite. London, 1893; reprint, York Beach, Maine, 1984.

Rupecissa, John of. *La Vertue et propriété de la quinte essence.* Lyon, 1549.

Sheppard, H. S. "The Redemption Theme and Hellenistic Alchemy." *Ambix* 7 (1959): 42–6.

Stillman, John Maxon. *The Story of Alchemy and Early Chemistry.* New York, 1960.

Stolcius de Stolcenberg, D. *Viridarium Chymicum Figuris, Cupro Incisis Adornatum, Et Poeticis picturis illustratum.* Frankfurt, 1624.

Taylor, F. Sherwood. *The Alchemists: Founders of Modern Chemistry.* New York, 1949.

Theatrum chemicum. 6 vols. Ursel, 1602.

Theatrum chemicum Britannicum. Ed. Elias Ashmole. London, 1562.

Thorndike, Lynn. "Alchemy During the First Half of the 16th Century." *Ambix* 2 (1938): 26–37.

——————. *History of Magic and Experimental Science up to the Seventeenth Century.* 8 vols. New York, 1929–58.

La Tourbe des philosophes. Paris, 1683.

Trismosin, Salmon. *Aureum vellus oder Guldin Schatz und Kunstkammer.* Rorschach am Bodensee, 1598–9.

——————. *La Toyson d'or.* Paris, 1613.

Valentinus, Basilius. *Chymische Schrifften.* Hamburg, 1700.

——————. *Triumphant Chariot of Antimony.* London, 1678.

Vallet de Viriville, Auguste. "Quelques recherches sur Nicolas Flamel." *Revue Français* 3 (1837).

_____. "Des ouvrages alchimiques attribués à Flamel." *Mémoire de la Société impériale des Antiquaires de France.* 3 (1857).

Vaughan, Thomas. *Works.* Ed. Alan Rudrum. Oxford, 1984.

Villain, Etienne F. *Essai d'une histoire de la paroisse Saint Jacques de la Boucherie.* Paris, 1758.

_____. *Histoire critique de Nicolas Flamel et de Pernelle sa femme, recueillie d'actes anciens qui justifient l'origine et la médiocrité du leur fortune contre les imputations des Alchimistes.* Paris, 1761.

Waite, Arthur Edward. *Alchemists Through the Ages.* 1888; reprint, New York, 1970.

_____, ed. *The Hermetic Museum Restored and Enlarged.* 2 vols. London, 1893.

Winternitz, Emanuel. *Musical Instruments and Their Symbolism in Western Art.* London, 1967.

Ziegler, Gilette. *Nicolas Flamel, ou le secret du Grand Oeuvre.* Paris, 1971.

Illustrated Manuscript Sources

Cambridge. Trinity College Library. MS Trinity O.8.24 (1399).

Leiden. Bibliotheek der Rijksuniversiteit. Cod. Voss. Chym. F. 29. Thomas Aquinas (pseud.). *De alchimia.*

London. British Library. MS Add. 1316. *Emblematical Figures of the Philosopher's Stone.*

_____. MS Add. 5025. *Ripley Scrowle.*

_____. MS Add. 32621. *Ripley Scrowle.*

_____. MS Harley 2407.

_____. MS Harley 3469. Salmon Trismosin, *Splendour solis* (1582)

_____. MS Sloane 288.

_____. MS Sloane 1520, Johannes Andreae.

London. Wellcome Institute for the History of Medicine Library. Wellcome MS 2456.

Munich. Deutches Museum. MS no. 38150.

Munich. Bayerische Staatsbibliothek. MS Germ. 598. *Book of the Holy Trinity.*

Nuremburg. Germanisches Nationalmuseum. MS 80061. *Book of the Holy Trinity.*

Oxford. Bodleian Library. MS 1566. *Ripley Scrowle.*

Paris. Bibliothèque de l'Arsenal. MS 6577. *Pretiosissimum donum Dei Georgium Anrach.*

Paris. Bibliothèque Nationale. MS Fr. 1089. *Glorieuse Marguerite.*

_____. MS Fr. 12297. Salmon Trismosin, *Splendor Solis.*

_____. MS Fr. 14765. Abraham le Juif. *Livre des figures hiéroglifiques.*

_____. MS grec. 2327. *Synosius.*

St. Gallen. Kantonsbibliothek Vadiana. MS 394a. *Rosarium philosophorum.*

_____. MS 428. *Book of the Holy Trinity.*

Vatican. Biblioteca Apostolica. MS Cod. lat. 7286. *Speculum veritatis.*

_____. MS Pal. Lat. 412. Wynandi de Stega, *Adamas colluctancium aquiliarum.*

Zurich. Zentralbibliothek. Cod. Rhen. 172. Thomas Aquinas (Pseudo), *Aurora consurgens.*

Sigla

1659: *Le livre figures hieroglyfiques de Nicolas Flamel.* . . . Ed. P. Arnauld de la Chevalerie. Paris, 1659, 1-139.

1692: *Nicholai Flammel Hieroglyphica.* In *Medicina practica.* Ed. William Salmon. London, 1692, 521-84.

OCD: *Oxford Classical Dictionary*

OED: *Oxford English Dictionary*

corr. Editor's correction.

Figure 1.   Fold-out Illustration of the tympanum of the Charnel House of the Innocents, Paris (now destroyed). Nicolas Flamel, *Exposition of the hieroglyphicall Figures which he caused to bee painted upon an Arch in St. Innocents Church-yeard in Paris.* London, 1624.

Figure 2. Tympanum Illustration showing the hieroglyphics of the *Book of Abraham* surrounding the arch. *Le Livre de Nicolas Flamel contenant l'explication des Figures hiéroglyphiques qu'il a fait mettre aucimetier des SS. Innocens à Paris.* In W. Salmon, ed., *Bibliothèque des philosophes (chimiques).* Paris, 1672.

Figure 3.   Snake Swallowing a Virgin.   Emblem 50.   Michael Maier, *Atalanta fugiens*. Oppenheim, 1618.

Figure 4. Alchemical Fountain. *Rosarium philosophorum*. Frankfurt, 1550.

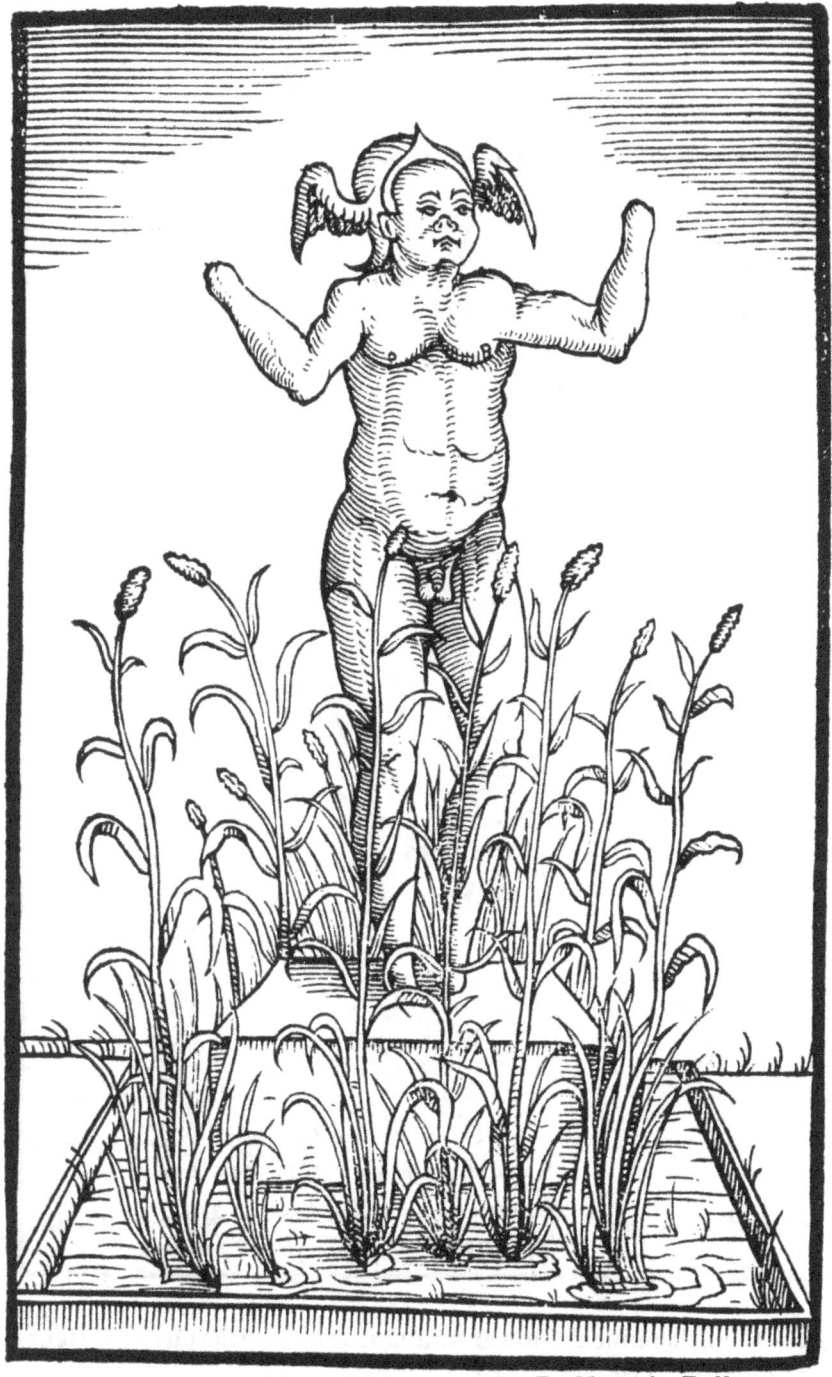

Figure 5.   The Fixation of Mercury. G. B. Nazari, *Della tramutatione metallica.* Brescia, 1599.

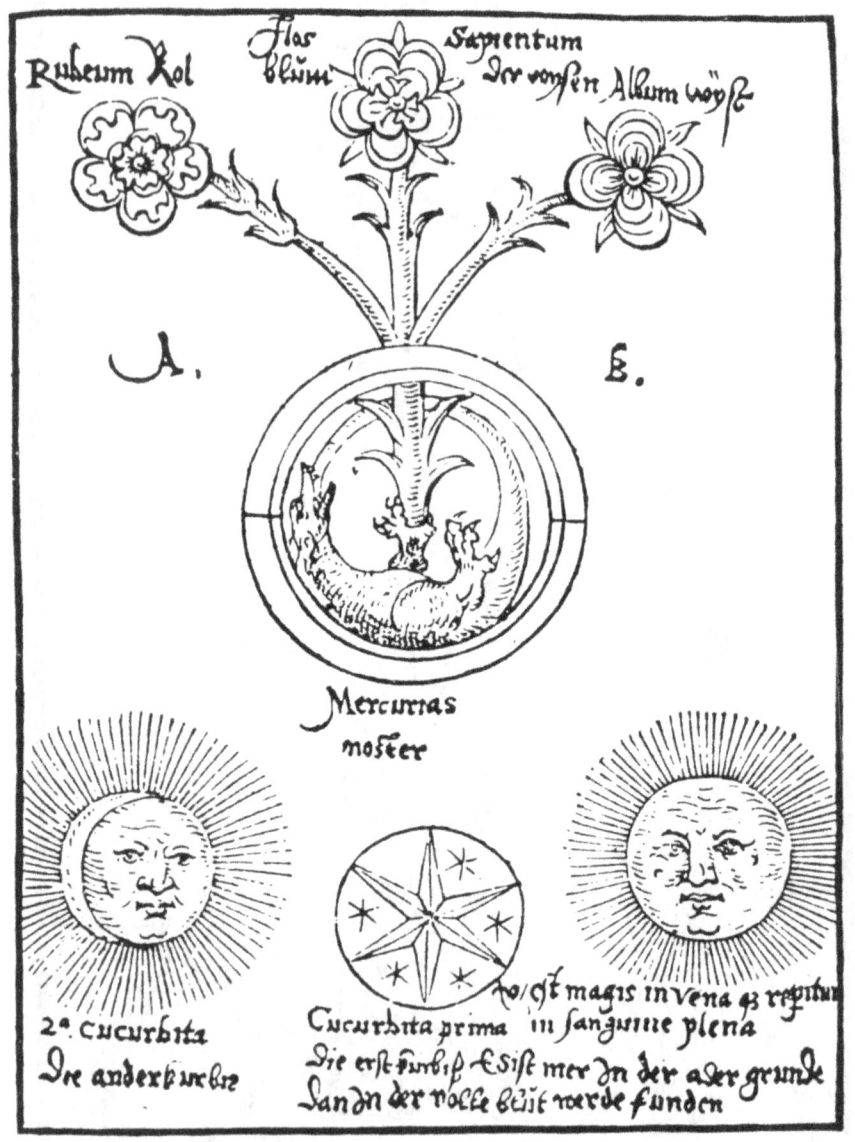

Figure 6.   Alchemical Flower and Ouroborus. Hieronymus Reusner, *Pandora*. Basel, 1588, 257.

Figure 7.   Alchemical Garden. Hieronymus Braunschweig, *Liber de arte distillandi*. Strasbourg, 1500. Frontispiece.

Figure 8. King and Queen Bathing. *Rosarium philosophorum.*
Frankfurt, 1550.

Figure 9.   Sun and Moon–King and Queen in the Bath. D. Mylius, *Philosophia reformata*. Frankfurt, 1662. No. 10.

Figure 10.   Battling the Dragon.   Lambsprinck, *De lapida philo-*
sophica. Frankfurt, 1625. Fig. 10.

Figure 11. Alchemical hermaphrodite. Reusner, *Pandora*. Basel, 1588.

Figure 12.   Christ as the *Lapis*. *Rosarium philosophorum*. Frank-furt, 1550.

Figure 13.   Alchemical sea arising from virgin's milk. Stolcius de Stolcenberg, *Viridarium chymicum*. Frankfurt, 1624.

Figure 14. Christ the judge and resuscitants rising from their graves. *Nicholai Flammel Hieroglyphica.* In W. Salmon, ed. *Medicinea practica.* London, 1692.

Nicholas Flammel,

His Exposition of the Hieroglyphicall
figures which he caused
to bee painted upon an Arch in
St. Innocents Church-yard,
in P A R I S.

Together with

The secret Booke of ARTEPHIUS,
And
The Epistle of *John Pontanus:*
Concerning both the Theoricke and
the Practicke of the
PHILOSOPHERS STONE.

Faithfully, and (as the Majesty of the thing
requireth) religiously done into English out
of the French and Latine Copies.

BY
IRENÆUS ORANDUS, *quiest,*
Vera veris enodans.

— ἀγαθῶν ἐπὶ δαῖτας ἴασιν
Αὐτόματοι ἀγαθοί.
[Plato *Symposium* 174b]

Imprinted at *London* by *T.S.* for *Thomas*
Walkley, and are to bee solde at his
Shop, at the Eagle and Childe in
Britans Bursse. 1624.

TO THE MOST

excellently accomplished

LADY, the *C.D.* of *E.*

MADAME:

Because there are not many worthy such Epithets, *therefore amongst so few, and those so dispersed, it is not hard for any man to know you, as well by your just* titles *as by your* Name. *Pardon my boldnesse, who owing my best service unto your vertues, though not knowing your person, nor knowne unto you, unlesse peradventure the report of my* disasters *have come unto your eares; doe humbly offer unto you, what I am assured, when you understand, (if ever God incline your* heart *to the* search, *and open your* eyes *to the* sight *thereof) you wil esteeme as the greatest and most unvaluable secret, which amongst all* under-Moone[1] *things, was ever imparted and communicated to man.* Your Piety and Almes deedes, *proceeding from that boundlesse fountaine of burning* Charity, *which disperseth it selfe in all formes, according to the necessities of the poore, have inforced mee to tell the world, that for you, and such as you are, I have caused these little* Bookes *to bee published in our* vulgar English, *custome excusing the most of your* sexe *from the knowledge of the* learned Tongues, *in which* Cabinets, *these* secrets *are ordinarily locked up, though there want not examples of many* women,[2] *who, by the impartiall grace of* God, *have attained to the thing it selfe. But it is not my purpose to flatter any body*

1

with the hope of that, which I well know how rare and reserved a blessing of the Almighty it is: Onely, if you will bee but pleased, by this occasion, to cast your eyes upon that triumphant Chariot, wherein Nature rideth through her Minerall and under-earth king-dome, you will easily see what difference there is, betweene the plenteous vertues of heaven, there thrust and crowded up together, (as lines though farre distant in their first setting foorth from the Circumference, yet touching one another when they come neere the Center) and the loose and weake composition of Vegetables, which being of another imposition of Nature, are not able either to receive or to hold such plentie of those heavenly Spirits, which are the life of every Elementary body, no where idle, and there most abound-ing where it seemes most to bee hidden. For the rest, if any of my busie unletter'd Countreymen, who are in great numbers, as bold pretenders to the blessed Science, as they are blinde practitioners therein, shall by the reading of these Treatises bee perswaded (as I wish they may) to forbeare the losse of their time, and the expence of their monyes, untill they be taught by the one of them, the true matter to worke on, and by the other, the true manner of proceed-ing therewith; let them in their hearts blesse God for you, to whose noble deserts (that chalenge a due acknowledgement from all good men) I have paid this small tribute of my labours. For mine owne part, the helpe and comfort which I have so plenteously reaped from these studies, in the middest of many pressures, which without the extra-ordinary assistance of God, had beene insupportable, hath already made light and easie in my resolution, whatsoever I shall either doe, or suffer, for God, or good men, or the trueth. The father of the fatherlesse, the Judge of the widdowes, and the hope of the helpelesse, bee to you and yours ALL THINGS. So prayeth,

Your humble servant

Eirenaeus Orandus.

E\t *sit splendor Domini Dei nostri super nos, & opera manuum nostrarum dirige super nos; & opus manuum nostrarum dirige. Psal.* 90.19.

And let the bright beauty of the Lord our God be upon us; and guide thou the workes of our hands upon us, and the work of our hands guide thou it. Psal.90.19.

Q\uis *enim despexit dies parvos? & laetabuntur, & videbunt lapidem stanneum in manu* Zorobabel. *Septemisti, Oculi sunt Domini, qui discurrunt in universam terram.* Zech. 4.10.

For who hath despised the day of little things? for they shall rejoyce, and shall see the stone of Tinne in the hand of *Zerubbabel*, with those seven; they are the eyes of the Lord, which run too & fro through the whole earth. Zech. 4.10.[3]

READER

— ἄλλα μὲν αὐτὸς ἐνὶ φρεσὶ σῇσι νοήσεις,
ἄλλα δὲ καὶ Δαίμων ὑποθήσεται. —

[Homer *Odyssey* 3.26-7]

Haec partim ipse tuo perpendes pectore tecum,
Partem Diuum aliquis tibi suggeret.

Part of these things thy mind shal prompt thee to,
And part, some God shall teach thee how to doe.

Againe.

Si te fata vocant, aliter non viribus ullis
Vincere, nec dure poteris convellere ferro.

If Fates thee call, else with no violence,
Nor hardest Iron canst thou dig them thence.[4]

Once againe, and so farewell.

Πολλαὶ μορφαὶ τῶν δαιμονίων,
Πολλὰ δ' ἀέλπτως χραίνουσι Θεοί,
καὶ τὰ δοκηθέντ' οὐκ ἐτελέσθη:
τῶν δ' ἀδοκήτων πόρον εὖρε Θεός.
τοιόν δ' ἀπέβη τόδε πρᾶγμα.

[Euripides *Alcestis* 1159-63]

Fortuna viccslul rica versat
Varias docilis sumere formas.
Inopina Dei plurima peragunt;
Non succedunt que fore speras,
Quae fore nemo posse putaret,
Sæpe expediunt numina. Qualem
Haec sortita est res mihi sinem.

Many shapes of Fate there bee
Much done beyond our hope, we see:
What we thinke sure, God often stayes,
And findes, for things undream't of, wayes.
For so did this succeed to mee,
And so I with it may to thee.

Eirenaeus Orandus.

5

THE BOOKE

of the

HIEROGLYPHICALL

Figures of

Nicholas Flammel.

Eternally praised be the Lord my God, which lifteth the humble from the base dust,[1] and maketh the hearts of such as hope in him to rejoyce: which of his grace openeth to them that beleeve, the Springs of his bountie, and putteth under their feet the worldly Sphæres (or circles)[2] of all earthly happinesses: In him bee alwayes our trust; in his feare, our felicitie; in his mercy, the glory of the reparation of our nature; and in our prayers, our unshaken assurance. And thou, o God Almighty, as thy benignity hath vouchsafed to open upon earth before me (thy unworthy servant) all the treasures of the riches of the world; so may it please thy great Clemencie, then when I shall be no more in the number of the living, to open unto me the treasures of heaven, and to let me behold thy Divine face, the Majestie whereof, is a delight unspeakeable, and the ravishing joy wherof, never ascended into the heart of living man. I aske it of thee, for our Lord Jesus Christ thy welbeloved Son his sake, who in the unity of the holy Spirit, liveth with thee world without end. Amen.

The Explication of the Hieroglyphicke Figures, placed by mee Nicholas Flammel, *Scrivener, in the Church-yard of the Innocents, in the fourth Arch, entring by the great gate of* St. Dennis *street, and taking the way on the right hand.*

The Introduction.

Although that I *Nicholas Flammel*, N O T A R Y, and abiding in *Paris*, in this yeere one thousand three hundred fourscore and nineteene,[3] and dwelling in my house in the street of Notaries, neere unto the Chappell of St. *James* of the *Bouchery*; although, I say, that I learned but a little Latine, because of the small meanes of my Parents, which neverthelesse were by them that envie me the most, accounted honest people; yet by the grace of God, and the intercession of the blessed Saints in *Paradise* of both sexes, and principally of saint *James* of *Gallicia*, I have not wanted the understanding of the Bookes of the *Philosophers*, and in them learned their so hidden secrets. And for this cause, there shall never bee any moment of my life, when I remember this high good, wherein upon my knees (if the place will give me leave) or otherwise, in my heart with all my affection, I shall not render thanks to this most benigne God, which never suffereth the child of the Just to beg from doore to doore, and deceiveth not them which wholly trust in his blessing.

Whilest therefore, I *Nicholas Flammel, Notary*, after the decease of my Parents, got my living in our Art of Writing, by making *inventories*, dressing accounts,[4] and summing up the Expences of *Tutors* and *Pupils*, there fell into my hands, for the sum of two Florens,[5] a guilded Booke, very old and large; It was not of Paper, nor Parchment, as other Bookes bee, but was onely made of delicate Rindes[6] (as it seemed unto me) of tender yong trees: The cover of it was of brasse, well bound, all engraven with letters, or strange figures; and for my part, I thinke they might well be *Greeke Characters*, or some such like ancient language: Sure I am, I could

7

not reade them, and I know well they were not notes nor letters of the *Latine* nor of the *Gaule*,[7] for of them wee understand a little. As for that which was within it, the leaves of barke or rinde, were ingraven, and with admirable diligence written, with a point of *Iron*,[8] in faire and neate Latine letters coloured. It contained thrice seven leaves, for so they were counted in the top of the leaves, and alwayes every seventh leafe was without any writing, but in stead thereof, upon the first seventh leafe, there was painted a *Virgin*, and *Serpents* swallowing her up; In the second seventh, a *Crosse* where a *Serpent* was crucified; and in the last seventh, there were painted *Desarts,* or *Wildernesses*, in the middest whereof ran many faire fountaines, from whence there issued out a number of *Serpents*, which ran up and downe here and there.[9] Upon the first of the leaves, was written in great Capitall Letters of gold, ABRAHAM THE JEW, PRINCE, PRIEST, LEVITE, ASTROLOGER, AND PHILOSOPHER, TO THE NATION OF THE JEWES, BY THE WRATH OF GOD DISPERSED AMONG THE GAULES, SENDETH HEALTH. After this it was filled with great execrations and curses (with this word MARANATHA,[10] which was often repeated there) against every person that should cast his eyes upon it, if hee were not *Sacrificer* or *Scribe*.

Hee that sold mee this Booke, knew not what it was worth, no more than I when I bought it; I beleeve it had beene stolne or taken from the miserable *Jewes*; or found hid in some part of the ancient place of their abode.[11] Within the Booke, in the second leafe, hee comforted his *Nation*, counceling them to flie vices, and above all, *Idolatry*, attending with sweete patience the comming of the *Messias*, which should vanquish all the Kings of the Earth, and should raigne with his people in glory eternally. Without doubt this had beene some very wise and understanding man. In the third leafe, and in all the other writings that followed, to helpe his *Captive nation* to pay their *tributes* unto the *Romane* Emperours, and to doe other things, which I will not speake of, he taught them in common words the *transmutation of Mettalls*; hee painted the *Vessels* by the sides,[12] and hee advertised them of the *colours*, and of all the rest, saving of the *first Agent*,[13] of the which hee spake not a word but onely (as hee said) in the fourth and fifth leaves entire hee painted it, and figured it with very great cunning and workemanship: for although it was well and intelligibly figured and painted, yet no man could ever have beene able to understand it, without being well skilled in their *Cabala*,[14] which goeth by tradition, and without having well studied their bookes. The fourth and fifth leafe therefore, was without any writing, all full of faire figures *enlightened*,[15] or as it were *enlightened*, for the worke was very exquisite. First he painted a *yong man*, with wings at his anckles, having in his hand a *Caducæan*[16] rodde, writhen about with two *Serpents*, wherewith hee strooke upon a helmet which covered his head; he seemed to my small judgement, to be the God *Mercury* of the *Pagans*: against him there came running and flying with open wings, a great old man, who upon his head had an *houre-glasse* fastened, and in his hands a hooke (or sithe) like *Death*, with the which, in terrible and furious manner, hee would have cut off the feet of *Mercury*.[17] On the other side of the fourth leafe, hee painted a faire *flowre* on the top of a very high *mountaine*, which was sore shaken with the *North wind*; it had the foot *blew*, the flowres *white* and *red*, the leaves shining like fine *gold*: And round about it the *Dragons* and *Griffons* of the *North* made their nests and abode. On the fifth leafe there was a faire *Rose-tree* flowred in the middest of a sweet *Garden*, climbing up against a hollow *Oake*; at the foot whereof boyled a fountaine of most *white water*, which ranne head-long downe into the depths, notwithstanding it first passed among the hands of infinite people, which digged in the Earth seeking for it; but because they were blinde, none of them knew it, except here and there one which considered the *weight*.[18]

On the last side of the fift leafe, there was a *King* with a great *Fauchion*,[19] who made to be killed in his presence by some *Souldiers* a great multitude of little *Infants*, whose Mothers wept at the feet of the unpittifull *Souldiers*: The bloud of which *Infants* was afterwards by other souldiers gathered up, and put in a great vessel, wherein the *Sunne* and the *Moone* came to bathe themselves.[20] And because that this History did represent the more part of that of the *Innocents* slaine by *Herod*, and that in this Booke I learned the greatest part of the *Art*, this was one of the causes, why I placed in their *Churchyard* these *Hieroglyphick*

Symbols of this secret science. And thus you see that which was in the first five leaves: I will not represent unto you that which was written in good and intelligible Latine in all the other written leaves, for God would punish me, because I should commit a greater wickednesse, then he who (as it is said) wished that all the men of the World had but one head that hee might cut it off at one blow.[21] Having with me therefore this *faire Booke*, I did nothing else day nor night, but study upon it, understanding very well all the operations that it shewed, but not knowing with what matter I should beginne,[22] which made me very heavy and sollitary, and caused me to fetch many a sigh. My wife *Perrenelle*, whom I loved as my selfe, and had lately married, was much astonished at this, comforting mee, and earnestly demanding, if shee could by any meanes deliver mee from this trouble: I could not possibly hold my tonge, but told her all, and shewed her this *faire Booke*, whereof at the same instant that shee saw it, shee became as much enamored as my selfe, taking extreame pleasure to behold the *faire cover, gravings,*[23] *images*, and *portraicts*, whereof notwithstanding shee understood as little as I: yet it was a great comfort to mee to talke with her, and to entertaine my selfe, what wee should doe to have the interpretation of them. In the end I caused to bee painted within my *Lodging*, as naturally as I could, all the figures and portraicts of the *fourth* and *fifth* leafe, which I shewed to the greatest Clerkes in *Paris*, who understood thereof no more then my selfe; I told them they were found in a Booke that taught the *Phylosophers stone*, but the greatest part of them made a mocke both of me, and of that blessed *Stone*, excepting one called *Master Anselme*, which was a *Licentiate* in *Physick*,[24] and studied hard in this *Science*: He had a great desire to have seene my Book, and there was nothing in the world, which he would not have done for a sight of it: but I always told him, that I had it not; onely I made him a large description of the *Method*. He told mee that the first portraict represented *Time*, which devoured all; and that according to the number of the *size* written leaves, there was required the space of *size* yeeres, to perfect the *stone*; and then he said, wee must turne the *glasse*, and seeth it no more. And when I told him that this was not painted, but onely to shew and teach the first *Agent*, (as was said in the Booke) hee answered me, that this

10

decoction for *sixe* yeeres space, was, as it were, a *second Agent*; and that certainly the *first Agent* was there painted, which was the *white and heavy water*, which without doubt was *Argent vive*,[25] which they could not *fixe*,[26] nor cut off his *feete*, that is to say, take away his *volatility* save by that long decoction in the purest bloud of young Infants;[27] for in that, this *Argent vive* being joined with *gold* and *silver*, was first turned with them into an *herb*[28] like that which was there painted, and afterwards by corruption, into *Serpents*; which *Serpents*[29] being then wholly dried, and decocted by fire, were reduced into powder of *gold*, which should be the *stone*. This was the cause, that during the space of *one and twenty yeeres*, I tryed a thousand broulleryes, yet never with *bloud*, for that was wicked and villanous: for I found in my Booke, that the *Philosophers* called *Bloud*, the minerall spirit, which is in the *Mettals*, principally in the *Sunne, Moone*, and *Mercury*, to the assembling whereof, I alwayes tended; yet these interpretations for the most part were more subtile then true. Not seeing therefore in my workes the *signes*, at the time written in my Booke, I was always to beginne againe. In the end having lost all hope of ever understanding those *figures*, for my last refuge, I made a vow to God, and St. *James* of *Gallicia*,[30] to demand the interpretation of them, at some *Jewish Priest*, in some *Synagogue* of *Spaine*: whereupon with the consent of *Perrenelle*, carrying with me the *Extract* of the *Pictures*, having taken the *Pilgrims* habit and staffe, in the same fashion as you may see me, without this same *Arch* in the *Church-yard*, in the which I put these *hyeroglyphicall figures*, where I have also set against the wall, on the one and the other side, a *Procession*, in which are represented by order all the colours of the *stone*, so as they come & goe, with this writing in French.

> *Moult plaist a Dieu procession,*
> *S'elle est faicte en devotion:* that is,

> *Much pleaseth God procession,*
> *If't be done in devotion.*

Which is as it were the beginning of King *Hercules* his Book, which entreateth of the colours of the *stone*, entituled *Iris*, or the *Rainebow*,[31] in these termes, *operis processio multum naturæ placet*, that is, *The procession of the worke is very pleasant unto Nature*: the which I have put there expresly for the great *Clerkes*, who shall understand the *Allusion*. In this same fashion, I say, I put my selfe upon my way and so much I did, that I arrived at *Montjoy*,[32] and afterwards at *Saint James*, where with great devotion I accomplished my vow. This done, in *Leon*[33] at my returne I met with a Merchant of *Boloyn*,[34] which made me knowne to a *Physician*, a *Jew* by Nation, and as then a *Christian*, dwelling in *Leon* aforesaid, who was very skilfull in sublime Sciences, called Master *Canches*.[35] As soone as I had showen him the figures of my *Extraict*, hee being ravished with great astonishment and joy, demanded of me incontinently, if I could tell him any newes of the *Booke*, from whence they were drawne? I answered him in *Latine* (wherein hee asked me the question) that I hoped to have some good newes of the *Book*, if any body could decipher unto me the *Enigmaes*: All at that instant transported with great Ardor and joy, hee began to decipher unto mee the beginning: But to be short, hee wel content to learn newes where this Book should be, and I to heare him speake; and certainly he had heard much discourse of the Booke, but (as he said) as of a thing which was beleeved to be utterly lost, we resolved of our voyage, and from *Leon* wee passed to *Oviedo*, and from thence to *Sanson*,[36] where wee put our selves to Sea to come into *France:* Our voyage had beene fortunate enough, & all ready, since we were entred into this Kingdome, he had most truly interpreted unto mee the greatest part of my figures, where even unto the very points and prickes, he found great *misteries,* which seemed unto mee wonderfull, when arriving at *Orleans*, this learned man fell extreamely sicke, being afflicted with excessive vomitings, which remained still with him of those he had suffered at Sea, and he was in such a continuall feare of my forsaking him, that hee could imagine nothing like unto it. And although I was alwayes by his side, yet would he incessantly call for mee, but in summe hee dyed, at the end of the *seventh* day of his sicknesse, by reason whereof I was much grieved, yet as well as I could, I caused him to be buried in the *Church* of the *holy Crosse* at *Orleans*, where hee yet resteth;[37] God have his soule, for hee dyed a good *Christian*: And surely, if I be not hindered by death, I will give unto that *Church* some *revenew*, to cause some *Masses* to bee said for his soule every day. He that would see the manner of my arrivall, and the joy of *Perenelle*, let him looke upon us two, in this *City of Paris*, upon the doore of the *Chappell of St. James* of the *Bouchery*, close by the one side of my *house*, where wee are both painted, my selfe giving thankes at the feet of *Saint James of Gallicia*, and *Perenelle* at the feet of St. *John*, whom shee had so often called upon. So it was, that by the grace of God, and the intercession of the happy and holy *Virgin*, and the blessed Saints *James* and *John*, I knew all that I desired, that is to say, The first *Principles*, yet not their first *preparation*, which is a thing most difficult, above all the things in he world: But in the end I had that also, after long errours of *three yeeres*, or thereabouts; during which time, I did nothing but study and labour, so as you may see me without this *Arch*, where I have placed my *Processions* against the two Pillars of it, under the feet of St. *James* and St. *John*, praying alwayes to God, with my Beades in my hand, reading attentively within a Booke, and poysing the words of the *Philosophers:* and afterwards trying and prooving the diverse operations, which I imagined to my selfe, by their onely words. Finally, I found that which I desired, which I also soone knew by the strong *sent* and *odour* thereof. Having this, I easily accomplished the *Mastery*, for knowing the *preparation* of the first *Agents*, and after following my Booke according to the *letter*, I could not have missed it, though I would. Then the first time that I made *projection*,[38] was upon *Mercurie*, whereof I turned halfe a pound, or thereabouts, into pure *Silver*, better than that of the *Mine*, as I my selfe assayed, and made others assay many times. This was upon a Munday, the 17 of *January* about noone,[39] in my house, *Perrenelle* only being present; in the yeer of the restoring of mankind, 1382. And afterwards, following always my Booke, from word to word, I made *projection* of the *red stone*[40] upon the like quantity of *Mercurie*, in the presence likewise of *Perrenelle* onely, in the same house, the *five and twentieth day* of *Aprill* following, the same yeere, about five a *clocke* in the *Evening*; which I transmuted truely into almost as much pure *Gold*, better assuredly than common *Golde*, more soft, and more plyable. I may speake it with truth, I have made it three times, with the helpe of *Perrenelle*, who understood it as well as I, because she helped mee in my operations, and without doubt, if shee would have enterprised to have done it alone, shee had attained to the end and perfection thereof. I had indeed enough when I had once done it, but I found exceeding great pleasure and delight, in seeing and contemplating the *Admirable workes of Nature*, within the *Vessels*. To signifie unto thee then, how I have done it *three times*, thou shalt see in this *Arch*, if thou have any skil to know them, three *furnaces*, like unto them which serve for our *opperations*: I was afraid a long time, that *Perenelle* could not hide the extreme joy of her felicitie, which I measured by mine owne, and lest shee should let fall some word amongst her kindred, of the great *treasures* which wee possessed: for extreme *joy* takes away the understanding, as well a great *heavinesse*; but the goodnesse of the most great God, had not onely filled mee with this blessing, to give mee a *wife* chaste and sage, for she was moreover, not onely capeable of reason, but also to doe all that was reasonable, and more discreet and secret, than ordinarily other women are. Above all, shee was exceeding *devout*, and therefore seeing her selfe without hope of children, and now well stricken in yeeres, shee began as I did, to thinke of God, and to give our selves to the workes of *mercy*. At that time when I wrote this *Commentarie*, in the yeere *one thousand foure hundred and thirteene*, in the end of the yeere, after the decease of my faithfull companion, which I shall lament all the dayes of my life: she and I had already founded, and endued with revenewes 14. *Hospitals* in this *Citie* of *Paris*, wee had new built from the ground *three Chappels*, we had inriched with great gifts and good rents, *seven Churches*, with many reparations in their *Church-yards*, besides that which we have done at *Boloigne*, which is not much lesse than that which wee have done heere. I wil not speake of the good which both of us have done to particular poore folkes, principally to *widdowes* and poore *Orphans*, whose names if I should tel, and how I did it, besides that my reward should be given mee in this World, I should likewise doe displeasure to those good persons, whom I pray God blesse, which I would not doe for any thing in the World. Building therefore these *Churches, Churchyards*, and *Hospitals* in this *City*, I resolved my selfe, to cause to be painted in the *fourth*

Arch of the Church-yard of the *Innocents*, as you enter in by the great gate in St. *Dennis street*, and taking the way on the right hand, the most true and essentiall markes of the *Arte*, yet under *vailes*, and *Hieroglyphicall covertures*, in imitation of those which are in the gilded Booke of *Abraham* the *Jew*, which may represent *two things*, according to the capacity and understanding of them that behold them: First, the *mysteries* of our future and undoubted *Resurrection*, at the day of Judgement, and comming of good *Jesus*, (whom may it please to have mercy upon us) a Historie which is well agreeing to a *Churchyard*. And secondly, they may signifie to them, which are skilled in Naturall *Philosophy*, all the principall and necessary operations of the *Maistery*. These *Hieroglyphicke figures* shall serve as two wayes to leade unto the heavenly life: the first and most open sence, teaching the sacred *Mysteries* of our salvation; (as I will shew heereafter) the other teaching every man, that hath any small understanding in the *Stone*, the lineary way of the worke; which being perfected by any one, the change of evill into good, takes away from him the roote of all sinne (which is *covetousnesse*) making him liberall, gentle, pious, religious, and fearing God, how evill soever hee was before, for from thence forward, hee is continually ravished, with the great grace and mercy which hee hath obtained from God, and with the profoundnesse of his Divine & admirable works. These are the reasons which have mooved mee to set these formes in this fashion, and in this place which is a *Churchyard*, to the end that if any man obtaine this inestimable good, to conquere this *rich golden Fleece*,[41] he may thinke with himselfe (as I did) not to keepe the *talent of God* digged in the *Earth*, buying Lands and Possessions, which are the vanities of this world: but rather to worke charitably towards his brethren, remembring himselfe that hee learned this *secret* amongst the *bones* of the *dead*,[42] in whose number hee shall shortly be found; and that after this life, hee must render an account, before a just and redoubtable *Judge*, which will censure even to an idle and vaine word. Let him therefore, which having well weighed my *words*, and well knowne and understood my *figures*, hath first gotten elsewhere the knowledge of the first *beginnings and Agents*, (for certainely in these *Figures* and *Commentaries*, he shall not finde any step or information thereof) perfect to the glory of God the *Maistery* of *Hermes*,[43] remembring himself of the *Church Catholike, Apostolike, and Romane*; and of all other *Churches, Churchyards,* and *Hospitals*; and above all, of the *Church* of the *Innocents* in this *Citie,* (in the *Churchyard* whereof hee shall have contemplated these true demonstrations) opening bounteously his purse, to them that are secretly poore, honest people desolate, weake women, widdowes, and forlorne orphanes. So be it.

CHAP. I.

Of the Theologicall Interpretations, which may be
given to these Hieroglyphickes, according to the
sence of mee the Authour.

I have given to this *Churchyard*, a *Charnell house*, which is right over against this fourth *Arch*, in the middest of the *Churchyard*, and against one of the Pillers of this *Charnell house*, I have made bee drawne with a coale, and grosely painted, *a man all blacke*,[44] which lookes straight upon these *Hieroglyphickes*, about whom there is written in *French; Je voy merveille done moult Je m'esbahi:* that is, *I see a marveile, whereat I am much amazed*: This, as also three *plates* of *Iron* and *Copper* gilt, on the *East, West*, and *South* of the *Arch*, where these *Hieroglyphickes* are, in the middest of the *Churchyard*, representing the holy *Passion* and *Resurrection* of the *Sonne* of *God*; this ought not to be otherwise interpreted, than according to the common *Theologicall* sence, saving that this *black man*, may as well proclaime it a wonder to see the admirable workes of God in the *transmutation* of *Mettals*, which is figured in these *Hieroglyphicks*, which he so attentively lookes upon, as to see buried so many *bodies*, which shall rise againe out of their Tombes at the fearful day of *judgement.* On the other part I doe not thinke it needfull to interpret in a *Theological* sence, that *vessell* of *Earth* on the right hand of these figures, within the which there is a *Pen* and *Inkhorne,* or rather a vessell of *Phylosophy*,[45] if thou take away the *strings*, and joyne the *Penner*[46] to the *Inkhorne*: nor the other two like it, which are on the two sides of the figures of *Saint Peter*, and *Saint Paul*, within one of the which, there is an *N.* which signifieth *Nicholas*, and within the other an *F.* which signifieth *Flammell.* For these ves-

17

sels signifie nothing else, but that in the like of them, I have done the *Maistery* three times. Moreover, he that will also beleeve, that I have put these vessels in forme of *Scutchions*,[47] to represent this *Pen* and *Inkhorne*, and the capitall letters of my *name*, let him beleeve it if he will, because both these interpretations are true.

Neither must you interpret in a Theological sence, that writing which followeth, in these termes, NICHOLAS FLAMMEL, ET PERRENELLE SA FEMME, that is, *Nicholas Flammel, and Perrenelle his wife*, in as much as that signifieth nothing, but that I and my wife have given that *Arche*.

As to the third, fourth, and fifth Tables following, by the sides whereof is written, COMMENT LES INNOCENTS FURENT OCCIS PAR LE COMMANDEMENT DU ROY HERODES, that is, *How the Innocents were killed by the commandement of King Herod*.[48] The *Theologicall* sence is well enough understood by the writing, we must onely speake of the rest, which is above.

The two *Dragons* united together the one within the other, of colour *blacke* and *blew*, in a field *sable*, that is to say, *blacke*, whereof the one hath the *wings* gilded, and the other hath none at all, are the *sinnes* which naturally are *enterchayned*,[49] for the one hath his *originall* and birth from another: Of them some may be easily *chafed* away, as they *come* easily, for they flie towards us every houre; and those which have no *wings*, can never be chased away, such as is the *sinne* against the *holy Ghost*. The *gold* which is in the *wings*, signifieth that the greatest part of sinnes commeth from the *unholy hunger* after *gold*; which makes so many people diligently to hearken from whence they may have it: and the colour *black* and *blew*, sheweth that these are the desires that come out of the darke pits of hell, which we ought wholly to flye from. These two *dragons* may also morally represent unto us the Legions of *evill spirits* which are always about us, and which will accuse us before the just Judge, at the fearefull day of judgement, which doe aske, nor seeke nothing else but to sift us.

The man and the woman which are next them, of an *orange colour*, upon a field of *azure* and *blew*, signifie that men and women ought not to have their hope in this World, for the *orange colour* intimates despaire, or the letting goe of hope, as here; and the colour *azure* and *blew*, upon the which they are painted, shewes us that we must thinke of heavenly things to come, and say as the roule of the man doth, HOMO VENI ET AD JUDICIUM DEI, that is, *Man must come to the judgement of God*, or as that of the *woman*, VERE ILLA DIES TERRIBILIS ERIT, that is, *That day will be terrible indeed*, to the end that keeping our selves from the *Dragons*, which are *sinnes*, God may shew mercy unto us.

Next after this, in a field of *Synople*, that is *greene*, are painted two men and one woman rising againe, of the which one comes out of a *Sepulchre*, the other two out of the *Earth*, all three of colour exceeding *white* and *pure*, lifting their hands towards their eyes, & their eyes towards Heaven on high: Above these three bodies there are two *Angels* sounding musicall Instruments, as if they had called these dead to the day of judgement; for over these two *Angels* is the figure of our Lord *Jesus Christ*, holding the world in his hand, upon whose head an *Angell* setteth a Crowne, assisted by two others, which say in their roules, *O pater Omnipotens, o Jesu bone*, that is, *O Father Almighty, o good Jesu.* On the right side of this *Saviour* is painted St. *Paul*, clothed with *white* & *yellow*, with a Sword, at whose feete there is a man clothed in a gowne of *orange colour*, in which there appeared pleights or folds of *blacke* and *white*, (which picture resembleth mee to the life) and demandeth pardon of his sinnes, holding his hands joined together, from betweene which proceed these words written in a roule, DELE MALA QUÆ FECI, that is to say, *Blot out the evils that I have done:* On the other side on the left hand, is *Saint Peter* with his Key, clothed in *reddish yellow*, holding his hand upon a woman clad in a gown of *orange colour*, which is on her knees, representing to the life *Perrenelle*, which holdeth her hands joyned together, having a rowle where is written, CHRISTE PRECOR ESTO PIUS, that is, *Christ I beseech thee be pittifull:* Behind whom there is an *Angell* on his knees, with a roule, that saith, SALVE DOMINE ANGELORUM, that is, *All haile thou Lord of Angels.* There is also another *Angel* on his knees behind my Image, on the same side that *S. Paul* is on, which likewise holdeth a roule, saying, O REX SEMPITERNE, that is, *O King everlasting.* All this is so cleere, according to the explication of the *Resurrection* and future judgement, that it may easily be fitted thereto. So it seemes this *Arch* was not painted for any other purpose, but to

19

represent this. And therefore we neede not stay any longer upon it, considering that the least and most ignorant, may well know how to give it this interpretation.

Next after the *three* that are rising againe, come two *Angels* more of an *Orange colour* upon a *blew field*, saying in their *rowles*, SURGITE MORTUI, VENITE AD JUDICIUM DOMINI MEI, that is, *Arise you dead, come to the Judgement of my Lord.* This also serves to the interpretation of the *Resurrection*: As also the last Figures following, which are, *A man red vermillion*, upon a field of *Violet colour*, who holdeth the foot of a winged *Lyon*, painted of *red vermillion* also, opening his throate, as it were to devoure the *man*: For one may say that this is the Figure of an unhappy sinner, who sleeping in a Lethargy of his corruption and vices, dieth without repentance and confession; who without doubt, in this terrible Day shall bee delivered to the *Devill*, heere painted in forme of a *red roaring Lyon*, which will swallow and devoure him.

CHAP. II.

*The interpretations Philosophicall, according to
the Maistery of Hermes.*

I desire with all my heart, that he who searcheth the secrets of the *Sages*, having in his Spirit passed over these *Idea's* of the life and resurrection to come, should first make his profit of them: And in the second place, that hee bee more advised than before, that hee sound and search the depth of my *Figures, colours,* and *rowles*; principally of my *rowles*, because that in this *Art* they speake not vulgarly. Afterward let him aske of himselfe, why the Figure of Saint *Paul* is on the right hand, in the place where the custome is to paint S. *Peter?* And on the other side that of Saint *Peter*, in the place of the figure of Saint *Paul?* Why the Figure of Saint *Paul* is clothed in colours *white* and *yellow* and that of S. *Peter* in *yellow* and *red?* Why also the *man* and the *woman* which are at the feet of these two *saints*, praying to *God*, as if it were at the Day of *judgement*, are apparrelled in divers colours, and not naked, or else nothing but bones, like them that are rising againe? Why in this Day of *Judgement* they have painted this *man* and this *woman* at the feet of the *Saints?* for they ought to have beene more low on *earth*, and not in *heaven*. Why also the two *Angels* in *Orange colour*, which say in their rowles, SURGITE MORTUI, VENITE AD JUDICIUM DOMINI MEI, that is, *Arise you dead, come unto the Judgement of my Lord*, are clad in this colour, and out of their place, for they ought to bee on high in heaven, with the two other which play upon the *instruments?* Why they have a field *Violet* and *blew?* but principally why their roule, which speaks to the dead, ends in the open throate of the *red and flying Lyon?* I would then, that after these, and many other questions which may justly bee made, opening wide the eyes of his spirit, he come to conclude, that all this, not having beene done without cause, there must bee represented under this *barke*, some great *secrets*, which hee ought to pray *God* to discover unto him. Having then brought his beliefe by degrees to this passe, I wish also that he would further beleeve, that these *figures* and *explications* are not made for them that have never seene the Bookes of the *Philosophers*, and who not knowing the *Mettallicke* principles, cannot bee named *Children* of this *Science*; for if they thinke to understand perfectly these *figures*, being ignorant of the *first Agent*, they will undoubtedly deceive themselves, and never bee able to know any thing at all. Let no man therefore blame me, if he doe not easily understand mee, for hee will be more blame-worthy than I, inasmuch as not being initiated into these sacred and secret interpretations of the *first Agent*, (which is the *key* opening the gates of all *Sciences*) he would notwithstanding, comprehend the most subtile conceptions of the *envious Philosophers*, which are not written but for them who already know these principles, which are never found in any booke, because they leave them unto *God*, who revealeth them to whom he please, or else causeth them to bee taught by the living voyce of a *Maister*, by *Cabalisticall* tradition, which hapeneth very seldome. Now then, *my Sonne*, let mee so call thee, both because I am now come to a great age, and also for that, it may be, thou art otherwise a *child* of this *knowledge*, (*God* inable thee to learne, and after to worke to his glory) Hearken unto mee then attentively, but passe no further if thou bee ignorant of the foresaid Principles.

This *Vessell* of *earth*, in this forme, is called by the *Philosophers*, their *triple Vessell*,[50] for within it, there is in the middest a Stage, or a floore, and upon that a dish or a platter full of lukewarm ashes, within the which is set the *Philosopicall Egge*,[51] that is, a viall of glasse full of *confections* of *Art* (as of the *scumme* of the *red Sea*, and the *fat* of the *Mercuriall winde:)*[52] which thou seest painted in forme of a *Penner and Inkehorne*. Now this Vessell of *earth* is open above, to put in the *dish* and the *viall*, under which by the open gate, is put in the *Philosophicall fire*, as thou knowest. So thou hast *three vessels*;[53] and the *threefold vessell*: The envious have called an *Athanor*,[54] a *sive, dung, Balneum Mariae*,[55] a *Furnace, a Sphaere, the greene Lyon*,[56] a *prison, a grave, a urinall, a phioll*,[57] and a *Bolts-head*:[58] I my selfe in my *Summarie* or *Abridgement of Philosophy*,[59] which I composed foure yeeres and two moneths past, in the end thereof named it the *house* and *habitation* of the *Poulet*,[60] and the *ashes* of the *Platter*, the *chaffe* of the *Poulet*; The common name is an *Oven*, which I should never have found, if *Abraham* the *Jew* had not painted it, together with the fire proportionable, wherein consists a great part of the secret. For it is as it were the *belly*,[61] or the *wombe*,[62] containing the true naturall heate to animate our *yong King*:[63] If this *fire* be not measured *Clibanically*, saith *Calid the Persian*,[64] *sonne of Jasichus*; If it be kindled with a sword, saith *Pithagoras:* If thou fire thy Vessell, saith *Morien*,[65] and makest it feele the heate of the fire, it will give thee a box on the eare, and burne his *flowres*[66] before they be risen from the depth of his *Marrow*, making them come out *red*, rather than *white*, and then thy worke is spoiled; as also if thou make too little fire, for then thou shalt never see the end, because of the coldnesse of the *natures*, which shall not have had motion sufficient to digest them together:

The heate then of thy *fire* in this vessell, shall be (as saith *Hermes* and *Rosinus*) according to the *Winter*;[67] or rather, as saith *Diomedes*, according to the heate of a *Bird*, which beginnes to flie so softly from the signe of *Aries* to that of *Cancer*:[68] for know that the Infant at the beginning is full of *cold flegme*,[69] and of *milke*, and that too vehement *heate* is an enemy of the *cold* and *moisture* of our *Embrion*, and that the two enemies, that is to

say, our two elements of *cold* and *heate* will never perfectly imbrace one another, but by little and little, having first long dwelt
together, in the middest of the temperate heate of their *bath*,[70]
and being changed by long decoction, into *Sulphur incombustible.*
Govern therefore sweetly with equality and proportion, thy proud
and haughty natures, for feare lest if thou favour one more then
another, they which naturally are enemies, doe grow angry against
thee through *Jelousy,* and dry *Choller,*[71] and make thee sigh for
it a long time after. Besides this, thou must entertain them in
this temperate heate perpetually, that is to say, night and day,
untill the time that *Winter,* the time of the *moisture* of the matters, be passed, because they make their peace, and joyne hands in
being heated together, whereas should these natures finde themselves but one onely half houre without *fire,* they would become
for ever irreconcileable. See therefore the reason why it is said
in the Book of the *seventy precepts, Looke that their heate continue indefatigably without ceasing, and that none of their dayes
bee forgotten.* And *Rasis,*[72] the haste, saith hee, *that brings with
it too much fire, is alwaies followed by the Divell, and Errour.
When the golden Bird,*[73] saith *Diamedes,*[74] *shall be come just to
Cancer, and that from thence it shall runne toward Libra,*[75] *then
thou maist augment the fire a little: And in like manner, when
this faire Bird, shall fly from Libra towards Capricorne,*[76] *which
is the desired Autumne, the time of harvest, and of the fruits that
are now ripe.*

CHAP. III.

*The two Dragons of colour
yellowish, blew, and black
like the field.*

Looke well upon these *two Dragons,*[77] for they are the true principles or beginnings of this *Phylosophy,* which the *Sages* have not dared to shew to their owne Children. Hee which is undermost, without wings, hee is the *fixed,* or the *male,* that which is uppermost, is the *volatile,* or the *female, blacke and obscure,* which goes about to get the domination for many monthes. The first is called *Sulphur,* or heat and drinesse, and the latter *Argent vive,*[78] or cold, and moisture. These are the *Sunne* and *Moone* of the *Mercurial* source, and *sulphurous originall,* which by continual fire are adorned with *royall* habiliments, that being united, and afterward changed into a *quintessence,* they may overcome every thing *Mettallick,* how solid hard and strong soever it bee. These are the *Serpents* and *Dragons* which the ancient *Aegyptians* have painted in a *Circle,* the *head* biting the *tayle,*[79] to signifie that they proceeded from one and the same thing, and that it alone was sufficient, and that in the turning and *circulation* thereof, it made it selfe perfect: These are the *Dragons* which the ancient Poets have fained did without sleeping keepe & watch the golden Apples of the Gardens of the Virgins *Hesperides.*[80] These are they upon whom *Jason*[81] in his adventure for the Golden Fleece, powred the broth or liquor prepared by the faire *Medea,* of the discourse of whom the Books of the *Phylosophers* are so full, that there is no *Phylosopher* that ever was, but he hath written of it, from the time of the truth-telling *Hermes Trismegistus, Orpheus, Pythagoras, Artephius, Morienus,* [82] and the other following, even unto my selfe. These are the *two Serpents,* given and sent by *Juno,* [83] (that is, the nature *Mettallicke*) the which the strong *Hercules,* that is to say, the sage and wise man must *strangle* in his *cradle,*[84] that is, overcome and kill them, to make them putrifie, corrupt, and ingender, at the beginning of his worke. These are the *two Serpents,* wrapped and twisted round about the *Caduceus* or rod of *Mercury,* with the which hee exerciseth his great power, and transformeth himselfe as he listeth. He, saith *Haly,* that shall kill the one, shall also kill the other, because the one cannot die but with his brother. These two then, (which *Avicen* calleth the *Corassene bitch* and the *Armenian dogge*)[85] these two I say, being put together in the vessell of the *Sepulcher,* doe bite one another cruelly, and by their great poyson, and furious rage, they never leave one another, from the moment that they have seized on one another (if the *cold* hinder them not) till both of them by their slavering venome, and mortall hurts, be all of a goare bloud, over all the parts of their bodies; and finally, killing one another, be stewed in their proper *venome*, which after their death, changeth them into living and *permanent water;*[86] before which time, they loose in their corruption and putrifaction, their first naturall formes, to take afterwards one onely new, more noble, and better forme. These are the two *Spermes, masculine* and *fœminine,* described at the beginning of my *Abridgement of Phylosophy,*[87] which are engendred (say *Rasis,*[88] *Avicen,*[89] and *Abraham* the *Jew* within the *Reynes,* [90] and entrails, and of the operations of the foure *Elements.* These are the radicall moysture of mettalls, *Sulphur,* and *Argent vive,* not vulgar, and such as are sold by the *Merchants* and *Apothecaries,* but those which give us those two faire & deare bodies which wee love so much. These two spermes, saith *Democritus,*[91] are not found upon the *earth* of the *living*: The same, saith *Avicen,* but he addeth, that they gather them from the dung, ordure, and rottennesse of the *Sunne* and *Moone.* O happy are they that know how to gather them; for of them they afterwards make a *triacle,*[92] which hath power over all griefes, maladies, sorrowes, infirmities, and weaknesses, and which sighteth puissantly against *death,* lengthening the life, according to the permission of *God,* even to the time determined, triumphing over the miseries of this world, and filling a man with the riches thereof. Of these two *Dragons* or Principles *Mettallicke,* I have said in my fore-alledged *Summarie,* that the Enemy would by his heate inflame his enemy, and that then if they take not heed, they should see in the ayre a venomous fume & a stinking, worse in flame, and in poyson, than the envenomed head of a *Serpent,* and *Babylonian Dragon.*[93] The cause why I have painted these two *Spermes* in the forme of *Dragons,* is because their stinch is exceeding great, and like the stinch of them, and the *exhalations* which arise within the glasse, are darke, *blacke, blew,* and *yellowish,*[94] (like as these two *Dragons* are painted) the force of which, and of the *bodies* dissolved, is so venomous, that truely there is not in the world a ranker *poyson;* for it is able by the force and stench thereof, to mortifie and kill every thing living: The *Philosopher* never feeles this *stinch,* if he breake not his vessels, but only he judgeth it to be such, by the sight, and the changing of *colours*, proceeding from the rottennesse of his confections.

These colours then signifie the *putrefaction* and *generation* which is given us, by the biting and dissolution of our *perfect bodies*, which dissolution proceedeth from externall heate ayding, and from the *Pontique*,[95] *fierienesse*, and admirable sharpe vertue of the poyson of our *Mercurie*, which maketh and resolveth into a pure cloud, that is, into impalpable powder, all that which it finds to resist it: So the heate working upon and against the *radicall, mettallicke, viscous*, or *oylie* moisture, ingendereth upon the subiect, *blacknesse*. For at the same time, the Matter is dissolved, is corrupted, groweth blacke, and conceiveth to ingender; for all *corruption* is *generation*, and therefore ought *blacknesse* to be much desired; for that is the *blacke saile* with the which the *Ship* of *Theseus*[96] came back victorious from *Crete*, which was the cause of the death of his *Father;* so must this father die, to the intent, that from the *ashes* of this *Phœnix* another may spring, and that the *sonne* may bee *King*. Assuredly hee that seeth not this *blackenesse* at the beginning of his operations, during the dayes of the *Stone*; what other colour soever he see, hee shall altogether fayle in the *Maisterie*, and can doe no more with that *Chaos*: for hee workes not well, if hee *putrifie* not; because if he doe not *putrifie*, hee doeth not *corrupt* nor *ingender*, and by consequent, the *Stone* cannot take *vegetative* life to increase and multiply. And in all truth, I tell thee againe, that though thou work upon the true matter, if at the beginning, after thou hast put thy *Confections* in the *Philosophers Egge*, that is to say, sometime after the fire have stirred them up, if then, I say, thou seest not his *head of the Crow*,[97] the *blacke* of the *blackest blacke*, thou must begin againe, for this fault is irreparable, and not to be amended; especially the *Orange colour*, or *halfe red*, is to be feared, for if at the beginning thou see that in thine *Egge*, without doubt, thou burnest, or hast burnt the *verdure* and livelinesse of thy *Stone*. The colour which thou must have, ought to bee intirely perfected in *Blacknesse*, like to that of these *Dragons* in the space of *fortie dayes*: Let them therefore which shall not have these essentiall markes, retire themselves betimes from their operations, that they may redeeme themselves from assured losse. Know also, and note it well, that in this Art it is but nothing to have this *blacknesse*, there is nothing more easie to come by: for from almost all things in the world, mixed with *moysture*, thou mayest have a *blacknesse*, by the fire: but thou must have a *blacknesse* which comes of the perfect *Mettallicke bodies*, which lasts a long space of time, and is not destroyed in lesse than *five moneths*, after the which followeth immediately the desired *whitenesse*. If thou hast this, thou hast enough, but not all. As for the colour *blewish* and *yellowish*, that signifieth that *Solution* and *Putrefaction* is not yet finished, and that the colours of our *Mercury* are not as yet well mingled, and rotten with the rest. Then this *blacknesse*, and these colours, teach plainly, that in this beginning the matter, and compound begins to rotte and dissolve into powder, lesse than the *Atomes* of the *Sunne*, the which afterwards are changed into *coator permanent*.[98] And this dissolution is by the envious *Philosophers* called *Death, Destruction,* and *Perdition*, because that the *natures* change their *forme*, and from hence are proceeded so many *Allegories* of *dead men, tombes* and *sepulchres*. Others have called it *Calcination, Denudation, Separation, Erituration,*[99] and *Assation*, because the *Confections* are changed and reduced into most small pieces and parts. Others have called it *Reduction into the first matter, Mollification, Extraction, Commixtion, Liquefaction, Conversion of Elements, Subtiliation, Division, Humation, Impastation,* and *Distillation,* because that the *Confections,* are melted, brought backe into seed, softned, and circulated within the glasse. Others have called it *Xir,*[100] or *Iris, Putrefaction, Corruption, Cymmerian darkenesse, a gulfe, Hell, Dragons, Generation, Ingression, Submersion, Complection, Conjunction,* and *Impregnation,* because that the matter is black & waterish, and that the natures are perfectly mingled, and hold one of another. For when the heate of the *Sunne* worketh upon them, they are changed first into *powder*, or fat and glutinous *water*, which feeling the heate, flyeth on high to the *Poulets* head,[101] with the *smoake*, that is to say, with the wind and ayre, from thence this water melted, and drawne out of the *confections*, goeth downe againe, and in descending reduceth, and resolveth, as much as it can, the rest of the *Aromatical confections*, alwayes doing so, untill the whole bee like a blacke broath somewhat fat.

29

Now you see, why they call this *sublimation,* and *volatization,* because it flyeth on high, and *Ascension* and *Descension,* because it mounteth, & descendeth within the glasse. A while after, the water beginneth to *thicken* and *coagulate* somewhat more, growing very *blacke,* like unto pitch, and finally comes the *Body* and *earth,* which the envious have called *Terra fœtida,* that is, *stinking earth:* for then because of the perfect *putrefaction,* which is as naturall as any other can be; this earth stincks, and gives a smell like the odour of *graves* filled with rottennesse, and with bodies as yet charged with their naturall moysture.[102] This *earth* was by *Hermes* called *Terra foliata,* or the *Earth of leaves,* yet his true & proper name is *Leton,*[103] which must afterward bee *whitened.* The Ancient Sages that were *Cabalists,* have described it in their *Metamorphoses,* under the History of the Serpent of *Mars,* which had devoured the companions of *Cadmus,*[104] who shew him, piercing him with his lance against a *hollow Oake. Note this Oake.*

CHAP. IIII.

Of the man and the woman clothed in a gowne
of Orange colour upon a field azure and blew,
and of their rowles.[105]

The *man* painted here doth expresly resemble *my selfe* to the naturall, as the *woman* doth lively figure *Perrenelle*: The cause why wee are painted to the life, is not particular to this purpose, for it needed but to represent a *male*, and a *female*, to which our two particular resemblance was not necessarily required, but it pleased the *Painter*[106] to put us there, just as hee hath done higher in this *Arch*, at the feet of the Figure of Saint *Paul* and Saint *Peter*, according to what wee were in our youth; as hee hath likewise done in other places, as over the *doore* of the *Chappell* of Saint *James* in the *Bouchery* neere to my house, although that for this last there is a particular cause as also over the doore of *Saincte Geneviesve de's Ardans*, where thou maist see me, I made then to bee painted heere two *bodies*, one of a *Male*, and another of a *Female*, to teach thee, that in this second operation, thou hast truely, but yet not perfectly, two *natures* conjoyned and married together, the *masculine* and the *Fœminine*; or rather the foure *Elements*; and that the foure naturall enemies, the *hote* and *cold*, *dry* and *moist*, begin to approach amiably one towards another, and by meanes of the *Mediators* and Peace-makers, lay downe by the little and little, the ancient enmity of the old *Chaos*.[107] Thou knowest well enough who these *Mediators* and Peace-makers are, betweene the *hote* and the *cold* there is *moisture*, for he is kinsman[108] and allyed to them both; to *hote* by his *heate*, and to *cold* by his *moisture*: And this is the reason, why to begin to make this peace, thou hast already in the precedent operation, converted all the *confections* into *water* by *dissolution*. And afterward thou hast made to *coagulate* the *water*, which is turned into this *Earth*, *blacke* of the *blacke* most *blacke*, wholly to accomplish this peace; for the *Earth*, which is *cold* and *dry*, finding himselfe of kindred and allyance with the *dry* and *moist*, which are enemies, will wholly appease and accord them. Doest thou not then consider a most perfect mixture of all the foure *Elements*, having first turned them into *water*, and now into *Earth?* I will also teach thee heereafter the other conversions, into *ayer* when it shall be all *white*, and into *fire*, when it shall bee of a most perfect *purple*. Then thou hast heere two *natures* married together, whereof the one hath conceived by the other, and by this *conception* it is turned into the body of the *Male*, and the *Male* into that of the *Female*; that is

32

to say, they are made one onely body, which is the *Androgyne*, or *Hermaphrodite* of the *Ancients*,[109] which they have also called otherwise, *the head of the Crow*, or *natures converted*.[110] In this fashion I paint them heere, because thou hast two natures reconciled, which (if they be guided and governed wisely) can forme an *Embrion* in the wombe of the *Vessell*,[111] and afterwards bring foorth a most puissant *King*, invincible and incorruptible, because it will bee an admirable *quintessence*.[112] Thus thou seest the principall and most necessary reason of this representation: The second cause (which is also well to bee noted) was because I must of necessitie paint *two bodies*, because in this operation it behooveth that thou *divide* that which hath beene *coagulated*, to give afterwards *nourishment*,[113] which is *milke* of *life*, to the little *Infant* when it is borne, which is endued (by the living God) with a *vegetable soule*.[114]

This is a secret most admirable and secret, which for want of understanding, it hath made fooles of all those that have sought it without finding it, and hath made every man wise, that beholds it with the eyes of his *body*, or of his *spirit*.

Thou must then make two parts and portions of this *Coagulated body*, the one of which shall serve for *Azoth*,[115] to wash and clense the other, which is called *Leton*, which must be whitened: He which is washed, is the *Serpent Python*, which (having taken his being from the corruption of the slime of the *Earth* gathered together by the waters of the *deluge*,[116] when all the confections were water) must be killed and overcome by the arrowes of the *God Apollo*,[117] by the *yellow Sunne*, that is to say, by *our fire*, equall to that of the *Sunne*.

He which *washeth*, or rather the *washings*, which must be continued with the other moity,[118] these are the *teeth* of that *Serpent*, which the sage workeman, the valiant *Theseus*,[119] wil sow in the same *Earth*, from whence there shall spring up armed *Souldiers*, which shal in the end discomfit themselves, suffering themselves by opposition to resolve into the same nature of the *Earth*, and the workman to beare away his deserved conquests. It is of this, that the *Phylosophers* have written so often, and so often repeated it, *It dissolves it selfe, it congeales it selfe, it makes it selfe blacke, it makes it selfe white, it kils it selfe, and it quickens it selfe.* I have made their field be painted *azure* and *blew*, to shew that I doe but now beginne to get out from the most *blacke blacknesse*; for the *azure* and *blew*, is one of the first colours, that the *darke woman* lets us see, that is to say, *moisture* giving place a little to *heate* and *drinesse*: The *man* and *woman* are almost all *orange-coloured*, to shew that our *Bodies*, (or our *body*, which the wise men here call *Rebis*) hath not as yet *digestion* enough, and that the *moisture* from whence comes the *blacke blew* and *azure*, is but halfe vanquished by the *drinesse*.

For when *drinesse* beares rule, all will be *white*, and when it fighteth with, or is equall to the *moisture*, all will be in part according to these present colours. The envious have also called these *confections* in this operation, *Nummus,*[120] *Ethelia,*[121] *Arena,*[122] *Boritis,*[123] *Corsufle,*[124] *Cambar,*[125] *Albar æris, Duenech,*[126] *Randeric,*[127] *Kukul,*[128] *Thabricis,*[129] *Ebisemech,*[130] *Izir,*[131] &c. which they have commanded to make *white.*

The *woman* hath a *white* circle in forme of a *rowle* round about her body, to shew thee, that *Rebis* will beginne to become white in that very fashion, beginning first at the *extremities*, round about this white *circle*. *Scala Phylosophoru(m)*, that is, the Booke entituled, *The Phylosophers Ladder,*[132] saith thus; *The signe of the first perfect whitenesse, is the manifestation of a certaine little circle of haire,*[133] *that is passing over the head, which will appeare on the sides of the vessels round about the matter, in a kind of a cierine*[134] *or yellowish colour.*

There is written in their Rowles, *Homo veniet ad judicium Dei*, that is, *Man shall come to the Judgement of God: Vere* (Saith the *woman) illa dies terribilis erit*, that is *Truly that will be a terrible day*. These are not passages of holy *Scripture*, but onely sayings which speake according to the *Theologicall* sence, of the *Judgement* to come, I have put them there, to serve my selfe of them towards him, that beholds onely the grosse outward, and most naturall *Artifice*, taking the interpretation thereof to concerne only the *Resurrection*; and also it may serve for them, that gathering together the *Parables* of the *Science*, take to them the eyes of *Lynceus,*[135] to pierce deeper then the *visible objects*. There is then, *Man shall come to the judgement of God: Certainly that day shall be terrible.* That is as if I should have said; It behoves that this come to the colour of *perfection*,[136] to bee judged & clensed from all his *blacknesse* and filth, and to be *spiritualized* and *whitened. Surely that day will be terrible*, yet certainly, as you shall find in the *Allegory of Aristeus*.[137] Horror holds us in prison by the space of *fourescore dayes*, in the darknesse of the *waters*, in the extreme heate of the *Summer*, and in the troubles of the *Sea*.[138] All which things ought first to passe, before our *King*[139] can become *white*, comming from death to life, to overcome afterwards all his enemies. To make thee understand yet somewhat better this *Albification*,[140] which is harder and more difficult then all the rest, (for till that time thou mayest erre at every steppe, but afterwards thou canst not, except thou break thy *vessels*). I have also made for thee this Table following.

CHAP. V.

The figure of a man, like that of Saint Paul,
cloathed with a robe white and yellow, bordered
with gold, holding a naked Sword, having at his
feet a man on his knees, clad in a robe of orange
colour, blacke and white, holding a roule.

Marke well this *man* in the forme of *Saint Paul*, cloathed in a robe entirely of a *yellowish white*. If thou consider him well, he turnes his body in such a *posture*, as shewes that he would take the *naked Sword*, either to cut off the *head*, or to doe some other thing, to that *man* which is on his knees at his feete, cloathed in a robe of *orange colour*, *white* and *blacke*, which saith in his roule, DELE MALA QUÆ FECI, that is, *Blot out all the evill which I have done*; as if hee should say, TOLLE NIGREDINEM, *Take away from me my blacknesse*; A term of Art: for *Evill* signifieth in the *Allegory, Blacknesse*, as it is often found in *Turba Phylosophorum*:[141] *Seethe*[142] *it untill it come to blackenesse, which will be thought Evill*. But wouldest thou know what is meant by this *man*, that taketh the *Sword?* It signifies that thou must cut off the head of the *Crow*,[143] that is to say, of the man cloathed in divers *Colours*, which is on his knees. I have taken this pourtraict and figure out of *Hermes Trismegistus*, in his Booke of the *Secret Art*, where he saith, *Take away the head of this blacke man, cut off the head of the Crow*, that is to say, *Whiten our blacke*. *Lambspringk*[144] that noble *Germane*, hath also used it in the *Commentary* of his *Hieroglyphicks*, saying, *In this wood there is a Beast all covered with black, if any man cut off his head, he will loose his blacknesse, and put on a most white colour. Will you understand what that is? The blacknesse is called the head of the Crow,*

the which being taken away, at the instant comes the white colour:
Then that is to say, when the Cloud appeares no more, this body
is said to bee without an head. These are his proper words. In
the same sence, the *Sages* have also said in other places, *Take the*
Viper which is called, De rexa, cut off his *head,* &c. That is to
say, Take away from him his *blacknesse.* They have also used this
Periphrasis,[145] when to signifie the multiplication of the *Stone,*
they have fained a *Serpent Hydra,*[146] whereof, if one cut off one
head, there will spring in the place thereof ten; for the stone augments tenfold, every time that they cut off this *head of the Crow,*
that they make it *blacke,* and afterwards *white*; that is to say, that
they dissolve it anew, and afterward coagulate it againe.

Marke how this naked Sword is wreathed about with a *blacke*
girdle, and that the ends thereof are not so wreathed at all. This
naked shining *Sword,* is the stone for the *white* or the white stone,[147]
so often by the *Phylosophers* described under this forme. To come
then to this perfect and sparkling *whitenesse,* thou must understand the wreathings of this blacke girdle, and follow that which
they teach, which is the quantity of the imbibitions. The two ends
which are not wreathed about at all, represent the beginning and
the ending: for the beginning it teacheth that you must *imbibe*[148]
it at the first time gently and scarcely, giving it then a little milke,
as to a little *Child* new borne, to the intent that *Isir*, (as the *Au-*
thors say) be not drowned: The like must we doe at the end, when
wee see that our *King* is *full*, and will have no more. The middle of
these operations is painted by the five whole *wreathes,* or *rounds,*
of the *blacke girdle,* at what time (because our *Salamander*[149] lives
of the *fire*, and in the middest of the *fire*, and indeed is a *fire*, and
an *Argent vive,* or quicksilver, that runnes in the middest of the
fire, fearing nothing) thou must give him abundantly, in such sort
that the *Virgins milke*[150] compasse all the matter round about.

I have made to be painted blacke all these *wreaths* or rounds of
the girdle, because these are the *imbibitions,* and by consequent,
blacknesses: for the *fire* with the *moisture* (as it hath been often said) causeth *blacknesse.* And as these *five* whole *wreathes* or
rounds shew that you must doe this *five times* wholly, so likewise
they let you know, that you must doe this in *five* whole moneths, a

moneth to every *imbibition*: See here the reason why *Haly Aben-ragel* said, *The Coction or boiling of the things is done in three times fifty dayes:* It is true, that if thou count these little imbibitions at the beginning and at the end, there are seven. Whereupon one of the most envious hath said, *Our head of the Crow is leprous,*[151] *and therefore he that would clense it, hee must make it goe downe seven times into the River of regeneration of* Jordan, *as the Prophet commanded the leprous* Naaman *the Syrian.*[152] Comprehending herein the beginning, which is, but of a few dayes, the middle and the end, which is also very short. I have then given thee this Table, to tell thee that thou must *whiten* my body, which is upon the knees, and demandeth no other thing: for Nature alwayes tends to perfection, which thou shalt accomplish by the apposition of *Virgins milke*, and by the decoction of the matters which thou shalt make with this *milke*, which being dryed upon this body, wil colour it into this same *white yellow*, which he who takes the *Sword*, is clothed withall, in which colour thou must make thy *Corsufle*[153] to come. The vestments of the figure of *Saint Paul* are bordered largely with a *golden* and *red citrine* colour. *O my Sonne*, praise God, if ever thou seest this, for now hast thou obtained mercy from Heaven; *Imbibe* it then, and teine[154] it till such time as the little Infant be hardy and strong, to combate against the *water* and the *fire:* In accomplishing the which, thou shalt doe that which *Demagoras, Senior,*[155] and *Hali* have called, *The putting of the Mother into the Infants belly,*[156] which Infant the Mother had but lately brought forth*; for they call the *Mother*, the *Mercury of Phylosophers*, wherewith they make their imbibitions and fermentations, and the *Infant* they call the *Body*, to teine or colour the which this *Mercury* is gone out: Therefore I have given thee these two figures, to signifie the *Albification*; for in this place it is, that thou hast need of great helpe, for here all the World is deceived. This operation is indeed a *Labyrinth*, for here there present themselves a thousand wayes at the same instant, besides that, thou must goe to the *end* of it, directly contrary to the *beginning*, in *coagulating* that which before thou *dissolvedst*, and in making *earth* that which before thou madest *water*. When thou hast made it *white*, then hast thou overcome the *enchanted Bulles*, that cast fire and smoake out of their nostrils. *Hercules* hath clensed the *stable*[157] full of ordure, of rottennesse, and of *blackenesse*. *Jason* hath powred the decoction or broath, upon the Dragons of *Colchos*, and thou hast in thy power the horne of *Amalthaea*,[158] which (although it bee *white*) may fill thee all the rest of thy life with glory, honour, and riches. To have the which, it hath behooved thee to fight valiantly, and in manner of an *Hercules*; for this *Achelous*, this moist river, is indewed with a most mighty force, besides that hee often transfigures himselfe from one forme to another: Thus hast thou done all, because the rest is without difficultie: These transfigurations are particularly described in the *Booke* of the *seven Egyptian seales*,[159] where it is said (as also by all *Authors*) that the *Stone, before it will wholly forsake his blackenesse, and become white in the fashion of a most shining marble, and of a naked flaming sword, will put on all the colours that thou canst possibly imagine, often will it melt, and often coagulate it selfe, and amidst these divers and contrary operations, (which the vegetable soule which is in it makes it performe at one and the same time) it will grow Citrine, greene, red, (but not of a true red) it will become yellow, blew, and orange colour, untill that being wholly overcome by drynesse and heate, all these infinite colours will end in this admirable Citrine whitenesse*, of the colour of Saint *Pauls* garments, which in a short time will become like the colour of the naked *sword*; afterwards by the meanes of a more strong and long decoction; it will take in the end a *red Citrine* colour, and afterward the perfect *redde* of the *vermillion*,[160] where it will repose it selfe for ever. I will not forget, by the way, to advertise thee, that the milke of the *Moone*, is not as the *Virgins milke* of the *Sunne*;[161] thinke then that the *inbibitions* of *whitenesse*, require a more *white* milke, than those of a *golden rednesse*; for in this passage I had thought I should have missed, and so I had done indeed had it not beene for *Abraham* the *Jew*; for this reason I have made to bee painted for thee, the Figure which taketh the naked sword, in the colour which is necessary for thee; for it is the figure of that which whiteneth.

CHAP. VI.

Upon a greene field, three resuscitants, or which rise againe, two men and one woman, altogether white: Two Angels beneath, and over the Angels the figure of our Saviour comming to judge the world, clothed with a robe which is perfectly Citrine white.

I have so made to bee painted for thee a field *vert*, [162] because that in this decoction the confections become *greene*, and keepe this colour longer than any other after the *blacke*. This *greenenesse* shewes particularly that our *Stone* hath a vegetable soule, and that by the Industrie of *Arte* it is turned into a true and pure *tree*, [163] to bud abundantly, and afterwards to bring foorth infinite little sprigs and branches. *O happy greene* (saith the *Rosary) which doest produce all things, without thee nothing can increase, vegetate, nor multiply.* The three *folke* rising againe, clothed in *sparkling white*, represent the *Body, Soule,* and *Spirit,* of our *white Stone.*[164] The *Philosophers* doe ordinarily use these termes of *Art* to hide the secret from evill men. They call the *Body* that *blacke earth*, obscure and darke, which wee make *white:* They call the *soule* the other halfe divided from the *Body*, which by the will of God, and power of nature, gives to the *body* by his inbibitions and fermentations a vegetable soule, that is to say, power and vertue to bud, encrease, multiply, and to become white, as a naked shining sword: They call the *Spirit*, the *tincture & drynesse*; which as a Spirit hath power to pierce all *Mettallick* things; I should be too tedious, if I should shew thee how good reason they had to lay alwayes and in all places, *Our Stone hath semblably to a man, a Body, Soule, and Spirit:* I would only that thou note well, that as a man indued with a *Body, Soule,* and *Spirit*, is notwithstanding but one; so likewise thou hast now, but one onely white confection, in the which

neverthelesse there are a *Body*, a *Soule*, and a *Spirit*, which are inseparably united. I could easily give very cleare comparisons and
expositions of this *Body, Soule*, and *Spirit*; but to explicate them,
I must of necessitie, speake things, which God reserves to reveale
unto them that fear and love him, and consequently ought not to
bee written. I have then made to bee painted heere, a *Body*, a
Soule, and a *Spirit*, all *white*, as if they were rising againe, to shew
thee, that the *Sun*, and *Moone*, and *Mercurie*, are raised againe
in this operation, that is to say, are made *Elements* of ayre, and
whitened: for wee have heretofore called the *Blacknesse, Death*;
and so continuing the *Metaphor*, wee may call *Whitenesse, Life*;
which commeth not, but with, and by a *Resurrection*: The *Body*,
to shew this more plainely, I have made to be painted lifting up
the stone of his tombe, wherein it was inclosed: The *Soule*, because it cannot bee put into the *earth*, it comes not out of a *tombe*,
but onely I have made it bee painted amogst the *Tombs*, seeking
its body, in forme of a *woman*, having her haire dischevelled; The
Spirit which likewise cannot bee put in a grave, I have made to bee
painted in fashion of a man comming out of the *earth*, not from
a Tombe. They are all white; so the *blacknesse*, that is, *death*,
is vanquished, and they being whitened, are from hence-forward
incorruptible. Now lift up thine eyes on high, and see our *King*
comming, crowned and raised againe, which hath overcome Death,
the darkenesses, and moistures; behold him in the forme wherein
our *Saviour* shall come, who shall eternally unite unto him all pure
and cleane soules, and will drive away all impurity and uncleannesse, as being unworthy to bee united to his *divine Body*. So by
comparison (but first asking leave of the *Catholicke, Apostolicke,
and Romane Church* to speake in this manner, and praying every
debonaire soule to permit me to use this similitude) see heere our
white *Elixir*,[165] which from hence-forward will inseparably unite
unto himselfe every pure *Mettallicke* nature, changing it into his
owne most fine *silvery* nature, rejecting all that is impure, strange,
and *Heterogeneall*, or of another kind. Blessed be God, which of
his goodnesse gives us grace to bee able to consider this sparckling
white, more perfect and shining than any compound nature, and
more noble next after the *immortall soule*, than any substance

43

having life, or not having life; for it is a *quintessence*, a most pure *silver*, that hath passed the *Coppell*,[166] and *is seven times refined*,[167] saith the royall Prophet *David*.

It is not needfull to interprete what the two *Angels* signifie, that play on Instruments over the heads of them which are raised againe: These are rather divine spirits, singing the mervailes of *God* in this miraculous operation, than Angels that call to judgement: To make an expresse difference betweene these and them, I have given the one of them a *Lute*, the other a *haultboy*,[168] but none of them *trumpets*, which yet are wont to be given to them that are to call us to *Judgement*. The like may be said of the three Angels, which are over the head of our *Saviour*, whereof the one crowneth him, and the other two assisting, say in their *Rowles*, O PATER OMNIPOTENS, O JESU BONE, that is, *O Almighty Father, O good Jesu*, in rendring unto him eternall thanks.

CHAP. VII.

Upon a field violet and blew,
two Angels of an Orange colour,
and their Rowles.

This *violet* and *blew* field sheweth, that being to passe from the *white Stone* to the *red*, thou must inbibe it with a little *virgins milke* of the *Sun*, and that these colours come out of the *Mercuriall* moysture which thou hast dried upon the *Stone*. In this operation of rubifying, although thou doe imbibe, thou shalt not have much *blacke*, but of *violet, blew*, and of the colour of the *Peacocks taile*.[169] For our *Stone* is so triumphant in *drynesse*, that as soone as thy *Mercury* toucheth it, the nature thereof rejoycing in his like nature, it is joyned unto it, and drinketh it greedily, and therefore the blacke that comes of moysture, can shew it selfe but a little, and that under these colours *violet* and *blew*, because that *drynesse* (as is said) doth by and by governe absolutely. I have also made to be painted for thee, these two *Angels* with wings, to represent unto thee, that the two substances of thy *confections*, the *Mercuriall*. and the *sulphurous* substance, the *fixed* as well as the *volatile*, being perfectly fixed together, do also flie together within thy vessell: for in this operation, the fixed body wil gently mount to heaven, being all *spirituall*, and from thence it will descend unto the *earth*, and whethersoever thou wilt, following every where the *Spirit*, which is alwayes mooved upon the *fire*: Inasmuch as they are made one selfesame nature, and the compound is all *spirituall*, and the *spirituall* all *corporall*, so much hath it beene subtilized upon *our Marble*, by the precedent operations. The natures then are heere transmuted into *Angels*, that is to say, are made *spirituall* and most subtle, so are they now the true *tinctures*. Now remember thee to begin the *rubifying*, by the apposition of *Mercury Citrine red*, but thou must not powre on much, and onely once or twice, according as thou shalt see occasion; for this operation ought to bee done by a *dry fire*, and by a *dry sublimation* and *calcination*. And truely I tell thee heere a secret which thou shalt very seldome finde written, so farre am I from being envious, that would to God every man knew how to make *gold* to his owne will, that they might live, and leade foorth to pasture their faire flocks, without Usury or going to Law, in imitation of the holy *Patriarkes*, using onely (as our first Fathers did) to exchange one thing for another; and yet to have that, they must labour as well as now. Howbeit for feare to offend *God*, and to be the instrument of such a change, which peradventure would proove evill, I must take heed to represent or write where it is that wee hide the *keyes*, which can open all the doores of the secrets of nature, or to open or cast up the *earth* in that place, contenting my selfe to shew the things which will teach every one to whom *God* shall give permission to know, what property the signe of the *Balance* or *Libra* hath, when it is inlightened by the *Sunne* and *Mercury* in the moneth of *October*. These *Angels* are painted of an *orange colour*, to let thee know, that thy white confections have beene a little more digested, or boyled, and that the *blacke* of the *violet* and *blew* hath beene already chased away by the *fire:* for this *orange colour* is compounded of the faire *golden Citrine red* (which thou hast so long waited for) and of the remainder of this *violet* and *blew*, which thou hast already in part, banished and undone. Furthermore this *orange colour* sheweth, that the natures are digested, and by little and little perfected by the grace of *God*. As for their Rowle, which saith, SURGITE MORTUI, VENITE AD JUDICIUM DOMINI MEI, that is, *Arise you dead, and come unto the judgement of God my Lord;* I have made it be put there, onely for the *Theologicall* sence, rather than any other: It ends in the throate of a *Lyon* which is all red, to teach that this operation must not bee discontinued untill they see the *true red purple*, wholly like unto the *Poppey* of the *Hermitage*,[170] and the *vermillion*[171] of the painted *Lyon*, saving for multiplying.

CHAP. VIII.

The figure of a man, like unto Saint Peter
cloathed in a robe Citrine red, holding a key
in his right hand, and laying his left hand
upon a woman, in an orange coloured robe,
which is on her knees at his feete,
holding a Rowle.

Looke upon this *woman* clothed in a robe of *orange colour*, which doth so naturally resemble *Perrenelle* as she was in her youth; Shee is painted in the fashion of a *suppliant* upon her knees, her hands joyned together, at the feete of a *man* which hath a *key* in his *right hand*, which heares her graciously, and afterwards stretcheth out his *left hand* upon her. Wouldest thou know what this meaneth? This is the *Stone*, which in this operation demandeth two things, of the *Mercury of the Sunne*, of the *Philosophers'* (painted under the forme of a *man*) that is to say *Multiplication*, and a more rich *Accoustrement*; which at this time it is needfull for her to obtaine, and therefore the man so laying his hand upon her shoulder accords & grants it unto her. But why have I made to bee painted a *woman?* I could as well have made to bee painted a *man*, as a *woman*, or an *Angell* rather, (for the whole natures are now spirituall and corporall, masculine and fœminine:) But I have rather chosen to cause paint a *woman*, to the end that thou mayest judge, that shee demaunds rather this, than any other thing, because these are the most naturall and proper desires of a woman. To shew further unto thee, that shee demandeth *Multiplication*, I have made paint the *man* unto whom shee addresseth her prayers in the forme of *Saint Peter*, holding a *key*, having power

49

to open and to shut, to binde and to loose; because the envious
Phylosophers have never spoken of *Multiplication*, but under these
common termes of *Art*, APERI, CLAUDE, SOLVE, LIGA, that is,
Open, shut, binde, loose; opening and *loosing*, they have called the
making of the *Body* (which is alwayes *hard* and *fixt*) *soft fluid*, and
running like water: To *shut* and to *bind*, is with them afterwards
by a more strong decoction to *coagulate* it, and to bring it backe
againe into the forme of a *body*.

It behoved mee then, in this place to represent a *man* with
a *key*,[172] to teach thee that thou must now *open* and *shut*, that
is to say, *Multiply* the budding and encreasing natures: for look
how often thou shalt dissolve and fixe, so often will these natures
multiply, in *quantity, quality,* and *vertue,* according to the multiplication of *ten*; comming from this number to an *hundred,* from
an *hundred* to a *thousand,* from a *thousand* to *ten thousand,* from
ten thousand to an *hundred thousand,* from an *hundred thousand,*
to a *million,* and from thence by the same operation to *Infinity,* as
I have done three times, praised be God. And when thy *Elixir* is so
brought unto *Infinity,* one *graine* thereof falling upon a quantity
of molten mettall as deepe and vaste as the *Ocean,* it will teine it,
and convert it into most perfect *mettall,* that is to say, into *silver*
or *gold,* according as it shall have been *imbibed* and *fermented,* ex-
pelling & driving out farre from himself, all the impure and strange
matter, which was joyned with the mettall in the first *coagulation*:
for this reason therefore have I made to bee painted a *Key* in the
hand of the *man,* which is in the forme of *Saint Peter,* to signifie
that the *stone* desireth to be *opened* and *shut* for *multiplication*;
and likewise to shew thee with what *Mercury* thou oughtest to doe
this, & when; I have given the man a garment *Citrine red,* and the
woman one of *orange* colour. Let this sufice, lest I transgresse
the silence of *Pythagoras,* to teach thee that the *woman,* that is,
our *stone,* asketh to have the rich Accoustrements and *colour* of
Saint Peter. Shee hath written in her Rowle, CHRISTE PRECOR
ESTO PIUS, that is, *Jesu Christ be pittifull unto mee,* as if shee
said, *Lord be good unto mee, and suffer not that hee that shal be
come thus farre, should spoile all with too much fire: It is true,
that from henceforward I shal no more feare mine enemies, and
that all fire shall be alike unto me, yet the vessell that containes*

me, is alwaies brittle and easie to be broken: for if they exalt the fire overmuch, it will cracke, and flying a pieces, will carry mee, and sow mee unfortunately amongst the ashes. Take heed therefore to thy fire in this place, and governe sweetly with patience, this admirable *quintessence*, for the fire must be augmented unto it, but not too much. And pray the soveraigne *Goodnesse*, that it will not suffer the evill spirits, which keepe the *Mines* and *Treasures*, to destroy thy worke, or to bewitch thy *sight*, when thou considerest these incomprehensible motions of this *Quintessence* within thy vessell.

CHAP. IX.

Upon a darke violet field, a man red purple,
holding the foote of a Lyon red as vermillion,
which hath wings, & it seemes would
ravish and carry away the man.[173]

This field *violet* and *darke*, tels us that the *stone* hath obtained by her full decoction, the faire *Garments*, that are wholly *Citrine and red,* which shee demanded of *Saint Peter*, who was cloathed therewith, and that her compleat and perfect *digestion* (signified by the entire *Citrinity*) hath made her leave her old robe of *orange colour.* The *vermilion red* colour of this *flying Lyon*,[174] like the pure & cleere *skarlet* in graine, which is of the true *Granadored*,[175] demonstrates that it is now accomplished in all right and equality. And that shee[176] is now like a *Lyon*, devouring every pure *mettallicke* nature, and changing it into her true substance, into true & pure *gold,* more fine then that of the *best mines.* Also shee now carrieth this man out of this vale of miseries, that is to say, out of the discommodities of *poverty & infirmity*, and with her wings gloriously lifts him up, out of the dead and standing waters of *Ægypt,* (which are the ordinary thoughts of mortall men) making him despise this life and the riches thereof, and causing him night and day to meditate on *God,* and his *Saints,* to dwell in the *Emperiall Heauen,* and to drinke the sweet springs of the Fountains of *everlasting hope.* Praised be *God* eternally, which hath given us grace to see this most fair & all-perfect *purple* colour; this pleasant colour[177] of the *wilde poppy* of the *Rocke*,[178] this *Tyrian*,[179] sparkling and flaming colour, which is incapable of *Alteration* or *change,* over which the *heaven* it selfe, nor his *Zodiacke* can have no more domination nor power, whose bright shining rayes, that dazle the eyes, seeme as though they did communicate unto a man some supercoelestiall thing, making him (when he beholds and knowes it) to be astonisht, to tremble, and to be afraid at the same time. *O Lord, give us grace to use it well, to the augmentation of the Faith, to the profit of our Soules, and to the encrease of the glory of this noble* REALME.

Amen.

FINIS.

Artephivs

HIS SECRET

BOOKE,

Of the blessed S T O N E
called the P H I L O-
S O P H E R S.

LONDON

Printed by *T.S.* for *Tho. Walkley*
and are to be sold at his Shop
at the Eagle and Childe
in Britans Burſſe.
1624.

ARTEPHIUS
HIS SECRET BOOKE,
Of the blessed STONE
called the PHILOSOPHERS.

LONDON

Printed by *T.S.* for *Tho. Walkley*
and are to be sold at his Shop
at the Eagle and Childe
in Britans Bursse.
1624.

THE PREFACE
To the READER, in
the French and Latine
Copies.

A mongst *all the other* Philosophers *(loving Reader) only our* Artephius *is not envious, as himself affirmeth of himselfe in many places, and therefore he layeth downe the whole* Art *in most open words in this* Treatise, *interpreting as farre as he may, the doubtfull speeches and* Sophismes *of others; Neverthelesse least he should give unto the wicked, ignorant, and evill men, occasion and meanes to doe hurt, hee hath a little vailed the truth in the* Principalls *of the* Science *under an Arteficiall Methode, sometimes affirming sometimes denying, and making as though hee often repeated one and the same thing, whereas in those repetitions hee alwayes changeth some words, seeming often to say the contrary of what hee had said before, willing to leave unto the judgement of the Reader, the way of* Trueth, Vertue, *and* true Working, *which if any man finde, let him give immortall thankes to* God *alone; but if hee see that hee walketh not in the right way, let him reade over this* Author *againe and againe, untill hee understand his meaning. So did the learned* John Pontanus,[1] *which saith in his epistle Printed in* Theatrum Chimicum: They err *(saith hee, speaking of them that labour in this* Arte) they have erred, and they will alwayes erre, because the Philosophers in their books have never set downe the proper Agent, except only one, which is called

56

Artephius, but hee speakes for himselfe; and if I had not read *Artephius*, and understood whereof hee spake, I had never come to the Complement of the worke: *Therefore reade this Booke, and reade it againe, untill thou understand his speech, and so obtaine thy desired end. It shall bee needlesse to speake any more concerning our* Authour; *It sufficeth that by the grace of God, and the use of this wonderfull* Quintessence,[2] *hee lived a thousand yeeres, as witnesseth* Roger Bacon,[3] *in his Booke* of the wonderfull workes of nature, *and also the most learned* Theophrastus Paracelsus,[4] *in his Booke of* long life: *Which terme of a thousand yeeres, none of the other* Philosophers, *no nor the Father of them,* Hermes *himselfe, was able to attaine unto. Looke therefore, whether peradventure this man have not understood the vertues of our* Stone, *and the manner how to use it, better than the rest. Howsoever it bee, use thou it and our labours, to the glory of God, and the profit of this* Kingdome.[5] *Farewell.*

ARTEPHIUS
HIS SECRET
BOOKE.

Antimony[6] is of the parts of *Saturne*,[7] and hath in every respect the nature thereof: so this *Saturnine Antimonie*[8] agrees with the *Sunne*, having in it selfe *Argent vive*,[9] wherein no mettall is drowned but *gold*; that is to say; Gold onely is drowned in *Antimoniall Saturnine Argent vive*, and without that *Argent vive*, no mettall can bee whitened: It whiteneth therefore *Leton*, that is, *Gold*,[10] and it reduceth a perfect *Body* into its first matter, that is, into *Sulphur* and *Argent vive* of a white colour, and shining more than glasse.[11] It dissolves I say, the perfect Body which is of his nature; for this water is friendly, and pleasant to the *Mettalls*, whitening the *Sunne*, because it contains a white *Argent vive*. And from hence thou mayest draw a great secret, to wit, that the water of *Saturnine Antimony* ought to be *Mercuriall* and *white* to the end that it may whiten the *Gold*, not burning it, but dissolving and afterwards congealing it to the forme of white *Creame*. Therefore, saith the *Philosopher*, that this water maketh the *Body* to bee *volatile*, because after it hath beene dissolved in this water, and cooled againe, it mounts aloft upon the surface of the water; *Take* (saith he) *gold* crude, foliated, laminated, or calcined with *Mercury*, and put it into our *Vinegre Antimoniall*,[12] *Saturnine, Mercuriall,* and drawne from *Sal Ammoniack*[13] (as is said) in a broad vessell of glasse, foure fingers[14] high or more, and leave it therein a temperate heate; and in short time thou wilt see lifted up, as it were a liquor of oyle swimming aloft, in manner of a thinne skinne:[15] That gather with a spoone, or with a feather, dipping it in, and so doing many times in a day, untill there doe nothing more arise; aferward make the *water* vapour away by the fire,[16] that is to say, the superfluous humor of the *vinegre*, and there will remain unto thee a *fifth essence* of *Gold* in forme of a white oyle incombustible, wherein the *Phylosophers* have placed their greatest secrets; and this *oyle* is exceeding *sweete*, and is of great power to mitigate the pain and griefe of wounds.

All the secret then of this secret *Antimoniall*, is that by vertue thereof we know how to extract & draw out of the body of the *Magnesia*,[17] *Argent vive*,[18] not burning (and this is *Antimony* and *Mercuriall sublimate*) that is, we must draw a water living, incombustible, and then congeale it with the perfect *Body* of the *Sunne*, which is dissolved therein, into a nature and substance white, congealed as if it were creame, which maketh it all to become white: Neverthelesse, first of all this *Sunne* in his putrifaction and resolution in this water, in the beginning will loose his light, be darkened, & become *black*, and afterward will lift himselfe upon the water, and there will swimme upon it, by little and little a white colour in a white substance. And this is called to *whiten the red Leton*,[19] to sublime it *Phylosophically*, and to reduce it into his first matter, that is to say, into white *Sulphur* incombustible, and into *Argent vive* fixed; and so the terminated moisture, that is to say, *Gold*, our *Body*, by the reiteration of liquifaction in this our dissolving water, is turned and reduced into *Sulphur*, and *Argent vive* fixed: And so the perfect *Body* of the *Sunne* taketh life in this water, is revived, inspired, encreased, and multiplied in his kind, as all other things are; for in this water it commeth to passe, that the *Body* compounded of *two bodies*, of the *Sunne* and of the *Moone*,[20] puffeth up, swelleth, putrifieth as a graine of Corne,[21] becommeth great with young, is lifted up, and encreaseth, taking the substance & nature, living and vegetable.

Also our water, or our foresaid *vinegre*, is the *vinegre of Mountaines*, that is to say, of the *Sunne* and *Moone*, and therefore it is mixed with the *Sunne* and *Moon*, and cleaveth to them perpetually: to wit, the *Body* taketh from this *water* the tincture of *whitenesse*, and with it (the *water*) shineth with inestimable brightnesse. Hee therefore that knowes how to turne the *Body* into white *silver* medicinall,[22] hee may afterward by this white *gold*, easily turne all imperfect mettals into very good and fine silver. And this *white gold*, is by the *Phylosophers* called, their *white moone* the *white Argent vive* fixed, the *Gold* of *Alchimy*, and the *white smoake*. Therefore without that our *Antimoniall* vinegre, the *white gold* of *Alchimy*, cannot be made. And because in our *vinegre* there is a double substance of *Argent vive*, one of *Antimony*, and another of *Mercury* sublimed; it doth there give a double weight & substance of *Argent vive* fixed, and also augments therein *(in the gold)* the naturall colour, weight, substance, and tincture thereof.

Therefore our dissolving water, carries a great tincture and great fusion,[23] because that when it feeles the common fire, if there be in it the perfect *Body* of the *Sunne* or of the *Moone*, it suddenly maketh it to bee melted, and to be turned into his substance, *white* as it is, & addes colour, weight, and tincture to the *Body*. It hath also power to disolve[24] all things that may be melted, and it is a ponderous body, viscous, precious, and honourable, resolving all crude bodies into their first matter, that is, into *Earth*, & a viscous powder, that is to say, into *Sulphur* and *Argent vive*. If therefore thou put into this water any mettall,[25] filed, or attenuated, and leavest it for a time in a gentle heate, it will be all dissolved, and changed into a *viscous water*, or a *white oyle*, as is said. And so it molifies the *Body*, and prepares it to *fusion & liquifaction*, nay, it makes all things fusible, that is, stones and mettals, and afterwards gives them spirit and life. Therefore it disolves all things with a wonderful solution, turning the perfect *Body* into a fusible medicine, melting, penetrating, and more fixed, encreasing the weight and colour.

Worke therefore with it, and thou shalt obtaine from it that which thou desirest; for it is the *spirit* and the *soule* of the *Sunne* and the *Moone*, it is the *oyle*,[26] the dissolving *water*, the *fountaine*, the *Balneum Mariae*,[27] the *fire against Nature*, the *moist fire*, the *secret, hidden,* and invisible *fire*, and the most sharpe *vinegre*, of which a certaine ancient *Phylosopher* said, *I besought the Lord, and hee shewed me a certain cleane water, which I knew to be the pure vinegre, altering, piercing, and digesting.* The vinegre I say penetrative, and the instrument moving the *gold* or the *silver*, to putrifie, resolve, and to be reduced into his first matter, and it is the onely *Agent* in the whole World for this Art, that can resolve and reincrudate, or make raw againe the *Mettallicke Bodies*, with the conservation of their *species.* It is therefore the onely fit and natural mean, by which we ought to resolve the perfect *Bodies* of the *Sunne* and *Moone*, by an admirable and solemne dissolution, under the conservation of their *species*, and without any destruction, unlesse it be to a new, more noble, and better forme, or generation, that is to say, into the perfect *Stone* of the *Phylosophers*, which is their wonderfull, and hidden secret.

Now this water is a certain middle substance, cleere as pure *silver*, which ought to receive the tinctures of the *Sunne* and *Moone*, to the end that it may be congealed and converted into white and living *Earth*; for this water hath need of the perfect bodies, that with them after dissolution, it may bee congealed, fixed, and coagulated into *white Earth*; and their *solution* is also their *congelation*, for thay have one and the same operation, for the one is not dissolved, but that the other is congealed; neither is there any other water which can dissolve the *Bodies*, but that which abideth with them in matter and forme, nay, it cannot be permanent, except it bee of the nature of the other body, that they may be made one together. Therefore when thou seest the *water* coagulate it selfe with the *Bodies* that bee dissolved therein, rest assured that thy *Science, Methode,* and operations are true and *Phylosophicall*, and that thou proceedest aright in the *Art.*

Nature then *is amended in its like nature*; that is, *Gold* and *Silver* are amended[28] in our *water*, as our *water* also with the *Bodies*; which after is called the meane of the *Soule*, without the which wee can doe nothing in this *Art*; and it is the vegetable, animall, and minerall fire, preserving the fixed spirits of the *Sunne* and *Moone*, the destroyer and the Conquerour of *Bodies* because it destroyes, dissolves, and changeth *Bodies*, and mettallick formes, and makes them to bee no *Bodies*, but a fixed spirit, and turneth them into a moist, soft, and fluid substance, which hath ingression and power to enter into other imperfect *Bodies*, [29] and to be mixed with them by the smallest parts, and to colour them and make them perfect; which they could not doe when they were *Metallicke* bodies dry & hard, which have no entrance, nor power to colour and make perfect imperfect *Bodies*. And therefore to good purpose doe wee turne the *bodies* into a fluid substance, because every tincture will colour a thousand times more, when it is in a soft and liquid substance, then when it is in a dry one as appeares by *Saffron*: and consequently the transmutation of imperfect *Bodies*, is impossible to be done by perfect *Bodies*, while they are dry, except they bee first brought backe into their first matter, soft and fluid: from hence wee conclude, that we must make the *Moisture* returne, and so reveale that which is hidden; which is called the *reincrudation*, or the making raw againe of the *Bodies*, that is, the boyling and the softening them, untill they bee deprived of their hard and dry *corporality*, or bodilynesse; because that which is dry, doth not enter, nor colour any more then it selfe. Therefore the dry Earthly *Body* doth not teine, except it be teined, because as is above-said, that which is thicke and Earthy, entreth not, nor coloureth; and because it entreth not, therefore it alters not; wherefore *Gold* coloureth not, untill the hidden spirit be drawne from the belly thereof by our *white water*, and that it be made altogether a spirituall and *white fume*, the *white spirit*, and the *wonderfull soule*.

Wherefore wee ought by our water, to attenuate, alter, and soften the *perfect Bodies*, that they may afterward be mixed with the other *imperfect Bodies*: And therefore if wee had no other profit by that *Antimoniall water*, then this, that it makes the *Bodies* subtile, soft, and fluid, according to his owne nature, yet it were sufficient for us: for it brings backe the *Bodies* to their first originall of *Sulphur* and *Mercury*, that of these, we may afterwards in a short time, in lesse then *one houre* of the day, doe that above ground, which Nature wrought under ground in the mines of the Earth in a *thousand yeeres*, which is as it were miraculous. And therefore our finall secret, is by our water to make the *Bodies volatile, spirituall,* and teining water, which hath ingression or entrance into the other *Bodies*: for it makes the *Bodies* to be a very *spirit*, because it doth *incerate*, (that is, bring to the temper and consistence of waxe) the hard and dry *Bodies*, and prepares them to fusion, that is, turnes them into a permanent or abiding water. It makes then of the *Bodies* a most precious blessed *Oyle*, which is the true tincture, and the white *permanent water*, of nature hot & moist, temperate, subtile, and fusible as waxe, which pierceth, reacheth to the bottome, coloureth, & maketh perfect. Therefore our water doth incontinently dissolve *gold* and *silver*, and maketh them an incombustible Oyle, which may then be mixed with other imperfect *Bodies*: for our water turnes the *Bodies* into the nature of a fusible *salt*, which is by the *Phylosophers* called, *Sal Albroe*, [30] which is the best and the noblest of all salts, being in the regiment thereof fixed, and not flying the fire, and it is indeed an oyle, of nature hot, subtile, penetrating, reaching to the depth and entring, called the compleat *Elixir*, and it is the hidden secret of the wise *Alchimists*. Hee therefore that knoweth this *salt* of the *Sunne* and *Moone*, and the generation, or preparation thereof, and aferwards how to mixe it, and make it friendly to the other imperfect bodies; hee in truth knoweth one of the greatest secrets of Nature, and one way of perfection.

These *Bodies* thus dissolved by our *water*, are called *Argent vive*, which is not without *Sulphur*, nor *Sulphur* without the nature[31] of the *Luminaries* (or lights) because that the Lights (the *Sunne* and *Moone*) are the principall meanes, or middle things, in the forme, by which *Nature* passeth in the perfecting and accomplishing the generation thereof: And this *quick-silver*, is called the *Salt* honoured, and animated and pregnant, (or great with Childe) and *fire*, seeing that it is nothing but *fire*, nor *fire*, but *Sulphur*, nor *Sulphur*, but *quick-silver*, drawne from the *Sunne* and *Moon* by our water, and reduced to a stone of great price; that is to say, it is the matter of the *Lights*, altered from basenesse unto noblenesse. Note that this white Sulphur is the Father of *Mettals*, and their Mother together, it is our *Mercury*; and the *Minera* of *Gold*, and the *Soule*, and the *ferment*, and the minerall vertue, and the living *Body*, and the perfect *Medicine*, our *Sulphur*, and our *Quick-silver*, that is, *Sulphur* of *Sulphur*, and *Quick-silver* of *Quick-silver*, and *Mercury* of *Mercury*. The property therefore of our water is that it melteth *gold* and *silver*, and augments in them their native colour; for it turnes the *Bodies* from *Corporality*, into *Spirituality*, and this water it is which sends into the *Body* a white fume, which is the white soule, subtile, hot, and of much fierinesse. This water is also called the *bloudy stone*,[32] and it is the vertue of the *spirituall bloud*, without which nothing is done, & the subject of all liquable things, and of liquifaction, which agrees very well, and cleaveth to the *Sunne* and the *Moone*, neither is it ever separated from them, for it is of kinne to the *Sunne* and to the *Moone*, but more to the *Sun* then to the *Moone*; *Note this well:* It is also called the *mean* of conjoyning the tinctures of the *Sunne* and *Moone* with imperfect *Mettals*; for it turnes the *Bodies* into a true *tincture* to teine the other imperfect *Mettals*, and it is the water which *whiteneth*, as it is *white*, which quickeneth as it is a *soule*; and therefore (as the *Phylosopher* saith) soone entreth into its *body*. For it is a living water, which commeth to moisten its *earth*, that it may budde, and bring forth fruit in his time, as all things springing from the *Earth*, are engendred by the *dew* or *moisture*. The *Earth* therefore buddeth not without watring and moisture: It is the water of *May-dew*,[33] that clenseth the *Bodies*, that pierceth them like raine water, whiteneth them, and maketh *one* new *Body* of *two Bodies*. This water of life being rightly ordered with his *Body*, whiteneth it, & turneth it into his white colour; for the *water* is a white fume, and therefore the *Body* is whitened by it: *whiten the body then, and burne thy Bookes.* And betweene these two, that is, betweene the *Body* and the *water*, there is friendship, desire, and lust, as betweene the *male* and the *fœmale*, because of the neerenesse of their like natures: for our second living water is called *Azot*[34] washing the *Leton*, that is, the *Body*, compounded of the *Sunne* and *Moon* by our first water. This second water is also called the *soule* of our dissolved *Bodies*, of which *Bodies* wee have already tyed the *soules* together, to the end that they may serve the wise *Phylosophers*. O how perfect and magnificent is this *water*, for without it the worke could never bee brought to passe! It is also called the vessell of *Nature*, the belly, the wombe,[35] the receptacle of the tincture, the *Earth*, and the Nurse. It is the Fountaine in which the *King* and *Queene* wash themselves,[36] and the *Mother* which must be put and sealed in the belly of her *Infant*, that is, the *Sun* which proceeded from her, and which shee brought forth: and therefore they love one another as a *Mother* and a *Sonne*, and are easily joyned together, because they came from one & the same roote, and are the same substance and nature. And because this water is the water of the *vegetable* life, therefore it giveth *life*, and maketh the dead body to vegetate, encrease, & spring forth, and to rise from *death* to *life*, by *solution* and *sublimation*; and in so doing, the *Body* is turned into a *spirit*, and the *spirit* into a *body*, and then is made amity, peace, concord, and union between the contraries, that is, betweene the *Body* and the *spirit*, which reciprocally change their natures, which they receive and communicate to one another by the least parts, so that the *hot* is mixed with the *cold*, the *dry* with the *moist*, and the *hard* with the *soft*; and thus is there a mixture made of contrary natures, that is, of *cold* with *hot*, and of *moist* with *dry*, an admirable connexion & conjunction of enemies. Then our dissolution of *bodies*, which is made in this first water, is no other thing then a killing of the *moist* with the *dry*, because the *moist* is coagulated with the *dry*, for the moisture is contained, terminated, and coagulated into a *Body*, or into *Earth*, onely by *drinesse*. Let therefore the hard and dry *bodies* be put in our first water in a vessell well shut, where they may abide untill they be dissolved, and ascend on high, and then they may bee called a *new Body*, the *white gold of Alchimy*, the *white stone*, the *white Sulphur*, not burning, and the *stone* of *Paradice*, that is, the *stone* which converts imperfect *Mettals* into fine white silver: Having this, we have also the *Body*, *Soule*, and *Spirit*, all together, of the which *spirit* and *soule* it is said, that they cannot be drawn from the perfect *Bodies*, but by the conjunction of our dissolving water, because it is certaine that the thing *fixed*, cannot be lifted up, but by the conjunction of the thing *volatile*. The *spirit* then by the mediation of *water* and the *soule*, is drawne from the *Bodies*, and the *Body* is made no *Body*, because at the same instant the spirit with the soule of the *Bodies* mounteth on high into the upper part, which is the perfection of the *stone*, and is called *sublimation*.[37] This *sublimation* (saith *Florentius Catalanus*) is done by things sharpe, spirituall, and volatile, which are of a sulphurous and viscous nature, which dissolve the *Bodies*, and make them to be lifted up into the Ayre in the spirit. And in this *sublimation* a certain part and portion of our said first *water* ascendeth with the *Bodies*, joyning it selfe to them, ascending and subliming into a middle[38] substance, which holdeth of the nature of the two, that is, of the *Bodies*, and of the *water*; and therefore it is called the Corporall & spirituall compound, *Corsufle*,[39] *Cambar*,[40] *Ethelia*, [41] *Zandarach*,[42] the *good Duenech*, [43] but properly it is onely called the *water permanent*,[44] because it flyeth not in the fire, always adhering to the commixed *Bodies*, that is, to the *Sunne* and *Moone*, and communicating unto them a living tincture,[45] incombustible, and most firme, more noble and precious then the former which these *bodies* had, because from hence-forward this tincture can run as *oyle* upon the *bodies*, perforating and piercing with a wonderfull *fixion*, because this *Tincture* is the *spirit*, and the *spirit* is the *soule*, and the *soule* is the *body*, because in this operation the *body* is made a *spirit* of a most subtile nature, and likewise the *spirit* is *incorporated*, and is made of the nature of a *body* with *bodies*, and so our *stone* contains a *body*, a *soule*, and a *spirit*. *O Nature* how thou changest the *body* into a *spirit*, which thou couldst not doe, if the *spirit* were not incorporated with the *bodies*, and the *bodies* with the *spirits* made volatile, or flying, and afterward *permanent* or *abiding*. Therefore they have passed into one another, and are turned the one into the other by wisdome. *O wisdome*, how thou makest *Gold* to be *volatile* and fugitive, although by nature it be most *fixed*. It behoveth therefore to dissolve and melt these *Bodies* by our water, and to make them a permanent water, a *golden water* sublimed, leaving in the bottom the grosse, earthly, and superfluous dry. And in this sublimation the *fire* ought to be soft, and gentle; for if in this sublimation the *Bodies* bee not purified in a lent or slow fire, and the grosser earthly parts (*note well*) separated from the uncleannesse of the *dead*,[46] thou shalt be hindred from ever making thy worke perfect; for thou needest onely this subtile and light nature of the dissolved *Bodies*, which our water will easily give thee, if thou proceed with a slow fire, for it will separate the *Heterogeneall* (or that which is of another kinde) from the *Homogeneall*, (or that which is all of one kinde.)

Our compound therefore receiveth mundification or clensing by our *moist fire*, that is to say, dissolving and subliming that which is pure and *white*, and casting aside the *fœces*, [47] *like a voluntary vomit* (saith *Azinaban*). For in such a dissolution, and naturall sublimation, there is made a loosing, or an untying of the

Elements, a clensing and a separation of the pure from the impure, so that the pure and white ascendeth upward, and the impure and earthly fixed remaines in the bottome of the *water,* or the *vessell,* which must be taken away and remooved, because it is of no value, taking onely the middle *white substance,* flowing and melting, and leaving the *fœculent earth,* which remained below in the bottome, which came principally from the water, and is the drosse, and the *damned earth,* which is nothing worth, nor can ever doe any good, as doth the pure, cleare, white and cleane matter, which wee ought onely to take. And against this *Capharœan* rocke, the *ship* and knowledge of the *Schollers* and *students* in *Philosophy,* is often (as it happened also unto mee sometimes) most improvidently dashed and beaten, because the *Phylosophers* doe very often affirme the contrary, namely, that nothing must be remooved or taken away, but the moysture, that is, the *Blacknesse,* which notwithstanding they say and write, onely to deceive the unwise, grosse, and ignorant, which of themselves without a *Maister,* unwearied *reading,* or *Prayer* unto *God* Almighty, would like conquerours carry away this golden fleece.[48]

Note therefore, that this separation, division, and sublimation, is without doubt the *key* of the whole worke. After the putrifaction then, and dissolution of these *Bodies,* our *Bodies* doe lift themselves up to the surface of the dissolving water, in the colour of *whitenesse,* and this *whitenesse* is *life;* for in this *whitenesse,* the *Antimoniall* and *Mercuriall soule,* is by the appointment of nature, infused with the Spirits of the *Sunne & Moone,* which separateth the subtile from the thicke, and the pure from the impure, lifting up by little and little, the subtile part of the *Body,* from the dregs, untill all the pure be separated and lifted up: And in this is our *Philosophicall* and naturall sublimation fulfilled: And in this *whitenesse* is the soule infused into the *Body,* that is, the mineral vertue, which is more subtile than *fire,* being indeed the true quintessence and life, which desireth to bee borne, and to put off the grosse earthly *fœces,* which it hath taken from the *Menstrous* and corrupt place of his *Originall.*[49] And in this is our *Philosophicall* sublimation, not in the naughty common *Mercury,* which hath no qualities like unto them, wherewith our *Mercury* drawne from his *vitriolate* cavernes, is adorned. But let us returne to our *sub-*

limation. It is therefore most certaine in this *Art*, that this *soule* drawne from the *Bodies*, cannot be lifted up, but by the putting to of a volatile thing, which is of his owne kinde; by the which the *Bodies* are made *volatile* and spirituall, lifting up, subtiliating, and subliming themselves, against their owne proper nature, which is *bodily*, heavy and ponderous; and by this meanes they are made no *Bodies*, but incorporeall, and a *fifth essence*, of the nature of the *Spirit*, which is called *Hermes* his *Bird*,[50] and *Mercury* drawne from the *red* servant;[51] and so the earthy parts remaine below, or rather the grosser parts of the *Bodies*, which cannot by any wit or device of man be perfectly dissolved. And this *white fume*, this *white gold*, that is, this *quintessence*, is also called the compound *Magnesia*, which as a *man*, containes, or like a *man* is compounded of a *Body*, a *Soule*, and a *Spirit*: For the *Body* is the fixed *earth* of the *Sunne*, which is more than most fine, ponderously lifted up, by the force of our divine water; The soule is the tincture of the *Sunne* and of the *Moone*, proceeding from the conjunction or communication of these two: But the spirit is the minerall vertue of the two *Bodies*, and of the *water*, which carries the *soule*, or the white tincture upon the *Bodies*, and out of the *Bodies*, as the tincture of *Diers*,[52] is carried by water upon the *cloth*. And that *Mercuriall* spirit is the Bond or tyall of the soule of the *Sun*; And the *Body* of the *Sunne* is the *Body* of fiction, containing with the *Moone* the spirit and soule. The *spirit* therefore pierceth, the *body* fixeth, the *soule* coupleth, coloureth and whiteneth. Of these three united together, is our *Stone* made, that is, of the *Sunne*, and *Moone*, and *Mercury*. Then with our gilded (or golden) *water*, is extracted a nature surpassing all nature, and therefore except the *bodies* bee by this our water dissolved, imbibed, ground, softened, and sparingly and diligently governed, untill they leave their grossenesse and thicknesse; and be turned into a thinne and impalpable spirit, our labour will alwayes be in vaine, for unlesse the *bodies* bee changed into no *bodies*, that is, into the *Philosophers Mercury*, the rule of Art is not yet found, and the reason is, because it is impossible to draw out of the *bodies* that most thinne or subtile *soule*, which hath in it all tincture, if the *bodies* be not first dissolved in our water. Dissolve therefore the *bodies* in the *golden water*, and boyle them, untill by the water all the tincture come out into a *white* colour, or a *white oyle*, and when thou shalt see this whitenesse upon the *water*, then know that the *bodies* are dissolved or melted, and continue the decoction, untill they bring foorth the *cloude* which they haue conceived, darke, blacke, and white. Put therefore the perfect *bodies* in our water, in a vessell *Hermetically* sealed, upon a soft fire, and boyle them continually, untill they bee perfectly resolved into a most precious *oyle*: Boyle them (saith *Adsar*)[53] with a gentle fire, as it were for the hatching of *chickens*, untill the *bodies* bee dissolved, and their tincture most neerely conjoyned, (*marke well*) be wholly drawne out: for it is not drawne out all at once, but it commeth forth by little and little, every day and every houre, untill after a long time this dissolution be complete, & that which is dissolved do alwaies arise uppermost upon the *water*. And in this dissolution let the fire bee soft and continuall, untill the *bodies* bee loosed into a viscous impalpable water, and that the whole *tincture* come forth first in the colour of *blackness*, which is a signe of true *solution*: Then continue the decoction untill it become a *white permanent water*, for governing it in its bath,[54] it will afterward by cleare, and in the end become like common *argent vive*, climing thorow the ayre upon the *first water*. And therefore when thou seest the bodies dissolved into a *viscous* water, then know that they are turned into a *vapour*, and that thou hast the *soules* separated from the *dead bodies*, and by sublimation brought into the order and estate of *spirits*, whereupon both of them with a part of our *water,* are made *spirits*, flying and clyming into the *ayre*, and that there the *body* compounded of the *male* and *female*, of the *Sunne* and *Moone*, and of that most subtile nature, clensed by *sublimation*, taketh life, is inspired by his moysture, that is, by his *water*, as a man by the *Ayre*, and therefore from henceforth it will multiply, and increase in his kinde, like all other things. And therefore in such an *elevation* and *Philosophical* sublimation, they are all joyned one with another, and the new body, inspired by the *Ayre*, liveth vegetably, which is a wonder. Wherefore unlesse the *Bodies* bee subtilized and made thinne by *fire* and *water*, untill they doe arise like *spirits*, and bee made like water and fume, or like *Mercury*, there is nothing done in this *Arte*. But when they ascend, they are borne in the ayre, and changed in the ayre, and are made life with life,

69

in such sort that they can never bee separated, as *water* mixt with *water*. And therefore it is wisely said that the Stone is borne in the *Ayre*, because it is altogether *spirituall*; for *the vulture flying without wings, crieth upon the top of the mountaine, saying, I am the white of the blacke, and the red of the white, and the Citrine sonne of the red. I tell truth, and lie not.*[55]

It sufficeth thee therefore to put the *Bodies* in the vessell, and in the water once for all, and to shut the vessell diligently, untill a true separation be made, which by the envious is called *conjunction*,[56] *sublimation, assation*,[57] *extraction, putrefaction, ligation, despousation, subtiliation*,[58] *generation, etc.* and that the whole *Maistery* bee done. Do therefore as in the generation of a *man*, and every *vegetable*, put the seed once into the *wombe*, and shut it well. By this meanes thou seest that thou needest not many things, and that our worke requires no great charges, because there is but one *Stone*, one *Medicine*, one *Vessell*, one *Regiment*, and one successive *disposition* to the *white*, and to the *red*. And although we say in many places *take this*, and *take that*, yet wee understand that it behooveth to take but *one* thing, and put it *once* in the vessell, and to shut the vessell untill the worke be perfected; for these things are so set down by the envious *Philosophers*, to deceive the unwary, as is aforesaid. For is not this Art *Cabalisticall*, and full of secrets? And doest thou, foole, beleeve that wee doe openly teach the *secrets of secrets?* and doest thou take our words according to the literall sound? Know assuredly, (I am no whit envious as others are) he that takes the words of the other *Philosophers*, according to the ordinary signification and sound of them, hee doeth already, having lost *Ariadnes* thread,[59] wander in the middest of the *Laberinth*, and hath as good as appointed his money to perdition. But I, *Artephius*, after I had learned all the *Art* and perfect *Science* in the Bookes of the true-speaking *Hermes*, was sometimes envious, as all the rest, but when I had by the space of a *thousand* yeeres, or thereabouts (which are now passed over mee since my nativity, by the onely grace of God Almighty, and the use of this wonderfull *fifth essence*) when, I say, for so long time I had seene no man that could worke the *Maistery* of *Hermes*, by reason of the obscurity of the *Philosphers* words, mooved with pitie, and with the goodnesse becomming an honest man, I have determined in these last times of my life to write all things truely and sincerely that thou maist want or desire nothing to the perfecting of the *Philosophers Stone*, (excepting a certaine thing, which it is not lawfull for any person to say or to write, because it is alwayes revealed by *God*, or by a *Maister*, and yet in this Booke, he that is not stiffe-necked, shall with a little experience, easily learne it.) I have therefore in this Booke written the naked trueth, although cloathed with a few colours, that every good and wise man, may from this *Philosophicall* Tree happily gather the admirable Apples of the *Hesperides*.[60] Wherefore praised bee the most high *God*, which hath put this benignitie into our soule, and with a wonderfull long olde age, hath given us a true dilection of heart, wherewithall it seemeth unto mee, that I doe truely love, cherish, and imbrace all men. But let us returne unto the *Arte*. Surely our worke is quickly dispatched, for that which the heate of the *Sunne* doeth in a hundred yeeres in the Mines of the Earth for the generation of a *Mettall*, (as I have often seene) our *secret fire*, that is, our fierie *sulphureous water*, which is called *Balneum Mariae*,[61] worketh in short time.

And this work is no great labour to him that knoweth and understandeth it, neither is the matter so deare (considering a small quantity sufficeth) that it ought to cause any man to plucke backe his hand, because it is so short and easie, that it may well bee called the *worke of Women*,[62] and *the play of Children*.[63] *Work* then cheerefully (my sonne) *pray* to *God*, *read* Bookes continually, for one Booke openeth another, thinke of it profoundly; fly all things that vanish in the fire, for thou hast not thine intent in these combustible and consuming things, but onely in the decoction of thy water, drawne from thy lights. For by this water is colour and weight given infinitely, and this *water* is a *white fume*, which as a *soule* floweth in the perfect *bodies*, taking wholly from them their blacknesse and uncleannesse, and consoledating the *two bodies* into one, and multiplying their *water:* And there is no other thing that can take away their true colour from the perfect *Bodies*, that is, from the *Sunne* and *Moone*; but *Azoth*, that is, this our water, which coloureth and maketh white the *red Body*, according to the regiments thereof.

But let us speake of *fires*. Our *fire* therefore is *minerall, equall, continuall,* it vapours not, unlesse it be too much stirred up, it partakes of *sulphur,* it is taken otherwhere then from the *matter,* it pulleth downe all things, it dissolveth, congealeth, and calcineth, it is *artificiall* to finde, it is a short way (or an expence) without cost, at the least, without any great cost, it is *moist, vaporous, digestive, altering, piercing, subtle, ayery, non violent, not burning, compassing or environing, containing but one,* and it is the Fountaine of living water, which goeth about, and containeth the place where the *King* and *Queene* bathe themselves.[64] In all the worke this *moist fire* is sufficient for thee, at the beginning, middest, and end; for in it consisteth the whole *Art:* This is the fire *naturall, against nature, unnaturall,* and without burning; and finally, this fire is *hot, dry, moist,* and *cold,* thinke upon this, and work aright, taking nothing that is of a strange nature: And if thou doest not well understand these fires, hearken further to what I shall give thee, never as yet written in any Booke, from out of the abstruse and hidden cavilation of the Ancients, concerning *fires.*

We have properly *three fires,* without the which the *Art* cannot bee done, and hee that workes without them, takes a great deale of care in vaine. The first is the *fire* of the *Lampe,* which is *continuall, moist, vaporous, ayery,* and artificiall to finde; for the Lampe ought to bee proportioned to the closure (or enclosure) and herein wee must use great judgement, which commeth not to the knowledge of a workeman of a stiffe necke: for if the fire of the *Lampe* be not *geometrically* and duly proportioned and fitted to the *Furnace,* either for lacke of heate thou wilt not see the expected signes in their times, and so thou wilt loose thy hope by too long expectation, or else with too much heate thou wilt burne the flowers[65] of the *Gold,* and so sadly bewaile thy lost labour. The *second fire* is the fire of *ashes,* in which the vessell *hermetically* sealed is shut up; or rather it is that most gentle heate, which proceeding from the temperate vapour of the *lampe,* goeth equally round about the vessell: This *fire* is not *violent,* if it be not too much stirred up, it is digesting, altering, it is taken from another *Body* then the matter, it is but one, or alone, it is moist and innaturall, &c. The *third* is the naturall fire of our water, which for this cause is also called *fire against nature,* because it is *water;* and yet neverthelesse it makes a meere spirit of *gold,* which common fire cannot doe; this fire is minerall, equall, and partakes of *Sulphur*, it breakes, congeales, dissolves, and calcines[66] all, this is piercing, subtile, not burning, and it is the Fountaine of *living water*, wherein the *King* and *Queen* bathe themselues, whereof wee have neede in the whole worke, in the *beginning, middle,* and *ending,* but the other two abovesaid, wee doe not always need, but onely sometimes: Joyne therefore in the reading the Bookes of *Phylosophers* these *three* sorts of fire, and without doubt thou shalt understand all their cavillations concerning their fires.

As touching the *Colours,* hee that doth not make *blacke,* cannot make *white,* because *blacknesse* is the beginning of *whitenesse,* and a signe of putrifaction and alteration, and that the *Body* is now pierced and mortified. Therefore in the putrifaction in this water, there first appeares *blackenesse,* like unto the broth wherein bloud, or some bloudy thing is boyled. Secondly, the blacke *Earth* by continuall decoction is *whitened,* because the *soule* of the two *bodies* swimmes aloft upon the water like white creame; and in this onely *whitenesse,* all the spirits are so united, that they can never fly from one another. And therefore the *Leton* must be *whitened,* and teare the Bookes, least our hearts be broken, for this intire *whitenesse* is the true *stone* to the *white,* and the *body* ennobled by the necessity of his end, and the tincture of *whitenesse,* of a most exuberant reflexion, and shining brightnesse, which being mixed with a *Body,* never departeth from it. Here then note, that the *spirits* are not *fixed,* but in the *white* colour, which by consequent is more noble then the other colours, and ought more earnestly to be desired, considering it is, as it were, the complement & perfection of the whole worke. For our *Earth* is first putrified in *blacknesse,* then it is clensed in the elevation or lifting up, afterwards being dryed, the *blacknesse* departeth, and then it is *whitened,* and the darke moist dominion of the *woman*[67] perisheth, and then the white fume pierceth into the *new Body,* and the *spirits* are shut up, or bound together, in drinesse, and that which is corrupting, deformed and *blacke* with moisture vanisheth, and then the *new Body* riseth againe, cleere, white, and immortall, getting the victory over al his enemies. And as heate working upon that which is *moist,* causeth or engendreth *blackenesse,* which is the *first* colour, so by decoction ever more and more, heate working upon that which is dry, begetteth *whitenesse*, which is the *second* colour; and afterward working upon that which is purely & perfectly dry, it causeth *citrinity* and *rednesse*; and so much concerning the *Colours*.

We must therefore understand, that the thing which hath the *head red* and *white*, the *feet white*, and afterwards *red*, and yet before that, the *eyes blacke*,[68] this onely thing is our maistery: dissolve then the *Sun* and the *Moone* in our dissolving water, which is familiar, friendly, and of the next nature unto them, which is likewise to them sweete and pleasant, and as it were a *wombe*, a *mother*, an *Originall*, the beginning and the end of life, and that is the reason why they are amended in this water, because *Nature rejoyceth in Nature, and Nature containes Nature*, and in true *Mariage* they are joyned together, and made one nature, one new body, raised up, and immortall. And thus we must joyne consanguinity with Consanguinity, and then these natures will meete, and follow one another, putrifie themselves, engender themselves, and make one another rejoyce, because *Nature* is governed by *Nature*, which is neerest and most friendly to it. Our water then (saith *Danthin*) is the most pleasant, faire, and cleere Fountaine, prepared onely for the *King & Queen*, whom it knoweth very well, and they know it; for it drawes them to it selfe, and they abide therein to wash themselves two or three *dayes*, that is, two or three *moneths*; and it maketh them young againe, & faire. And because the *Sunne* and *Moone* have their Originall from this water their Mother, therefore it behoveth that they enter againe into their Mothers wombe, that they may be borne againe, and be made more strong, more noble, and more valiant. And therefore if these doe not *die*, and be not turned into water, they remain alone, and *without fruite*; but if they *die*, and be resolved in our water, they bring fruit an *hundreth* fold; and from that very place, where it seemed they had lost what they were, from thence shall they appeare that which they were not before. Let therefore the *spirit* of our living water, be with great wit and subtilty fixed with the *Sunne* and the *Moone*, because they being turned into the nature of water, doe dye, & seeme like unto the dead; yet afterward being inspired from thence, they live, encrease, and multiply like all other *vegetable* things. It is enough then to dispose the matter sufficiently from *without*, for from *within*, it selfe doth work sufficiently to its owne perfection. For it hath in it selfe a certaine and inhærent *motion*, according to the true way, better then any order that can be imagined by man. And therefore doe thou onely *prepare*, and Nature will *perfect*; for if shee bee not hindered by the contrary, shee will not passe her owne certaine *motion*, as well to conceive, as to bring forth. Wherefore after the preparation of the matter, take heede onely least by too much fire thou make the *bath* too hot: *Secondly*, take heed least the *spirit* doe exhale, because it would hurt him that worketh, that is to say, it would destroy the worke, and cause many *infirmities*, that is, much sadnesse and anger. From this that hath beene spoken, is drawne this *Axiome*, to wit, *that by the course of nature, he doth not know the making of Mettals, that knoweth not the destruction of them*. It behoveth then, to joyne together them that are of kindred, for *Natures* doe finde their like *natures*, and being putrified, are mixed together, and mortifie themselves. It is necessary therefore to know this *corruption* and *generation*, and how the Natures doe imbrace one another, and are pacified in a *slow* fire, how Nature *rejoyceth* in Nature, and nature *retaines* nature, and turnes it into a white nature. After this, if thou wilt make it *red*, thou must boyle this *white*, in a dry continuall fire, untill it bee as *red* as *blood*, which will bee nothing else but *fire* and a true *tincture*: And so by a continuall dry fire, the *whitenesse* is changed, amended, perfected, made *Citrine*, and acquireth *rednesse*, a true *fixed* colour. And consequently by how much more this *red* is boyled, so much the more is it coloured, and made a tincture of perfect *rednesse*; Wherefore thou must with a *dry fire*, and a *dry calcination*, without any moysture, boyle this *compound*, untill it bee clothed with a most *red* colour, and then it will be a perfect *Elixir*.

If afterwards thou wilt *multiply* it, thou must againe resolve that *red* in a new dissolving water, and after by decoction *whiten* and *rubifie* it by the degrees of fire, reiterating the first regiment. Dissolve, congeale, reiterate, shutting, opening, and multiplying in *quantitie* and *qualitie* at thine owne pleasure: for by a new corruption and generation, there is againe brought in a new *motion*, and so we could never find an end, if we would alwayes worke by reiteration of *solution* and *coagulation*, by the meanes of our dissolving *water*, that is to say, dissolving and congealing, as is said in the first regiment. And so the vertue thereof is increased and multiplied in *quantitie* and *qualitie*, so that if in the first worke, one part of thy *Stone*, will teyne an *hundred*, in the *second* it will teyne a *thousand*, in the *third ten thousand*, and so by pursuing thy worke, thy projection will come into *infinitie*, teyning truly and perfectly, and fixedly, every quantitie, how great soever it bee, and so by a thing of an easie price, is added *colour*, and *vertue*, & *weight*. Therefore *our fire and Azoth are sufficient for thee*; boyle, boyle, reiterate, dissolve, congeale, and so continue according to thy will, multiplying it as much as thou wilt, and untill thy Medicine bee made fusible as *waxe*, and that it have the quantitie and vertue which thou desirest. Therefore all the accomplishment of the worke, or of our *second Stone*, *(note it well)* consisteth in this, that thou take the *perfect Body*, which thou must put in *our water*, in a house of *glasse*,[69] wel shut and stopped with *Cement*, lest the *ayre* get in, or the *moysture* inclosed get out; and there hold it in the digestion of a gentle heate, as if it were of a *bathe*, or the most temperate heate of *dung*, upon the which with the fire thou shalt continue the perfection of decoction, untill it bee putrified and resolved into *blacke*, and afterwards be lifted up, and sublimed by the water, that it may thereby bee cleansed from all *blacknesse* and darknesse, and that it may bee *whitened* and made *subtile*, untill it come to the utmost purity of sublimation, and at the last be made *volatile*, and *white*, within and without: for the *vulture flying in the Ayre without wings, cryeth that it might get upon the Mountaine, that is, upon the water*, upon the which the *white* Spirit is carried. Then continue a convenient *fire*, and that *spirit*, that is, the subtile substance of the *Body* and of *Mercury* will ascend upon the water, which quintessence is *whiter* than the *snow*; continue still, and in the end strengthen thy fire, untill all which is *spirituall* mount on high: for know well, that all that is cleare, pure, and spirituall, ascends on high in the *ayre*, in the forme of a *white* fume, which the *Philosophers* call, the *Virgins milke*.

It behooveth therefore, that (as *Sibill* said) *the Sonne of the Virgin bee exalted from the Earth*, and that the white *quintessence* after his resurrection bee lifted up towards the heavens,[70] and that the grosse and thicke remaine in the bottome of the vessell and of the *water*; for afterwards when the vessell is colde, thou shalt finde in the bottome thereof, the *fœces*, *blacke*, burnt, and combust, separate from the *spirit* and *white quintessence*, which dregs thou must cast away. In these times the *Argent vive*[71] raineth from our *ayre* vpon our new *earth*, which is called *Argent vive*, sublimed from the ayre, whereof is made a *water* viscous, cleane and *white*, which is the true *tincture* separated from all *blacke fœces*, and so our *brasse* or *Leton*, is with our water governed, purified, and adorned with a *white* colour, which *white* colour is not gotten, but by decoction and coagulation of the *water*. Boyle it then continually, wash away the *blacknesse* from the *Leton*, not with thy hand, but with the *Stone*, or the *fire*, or our *second Mercuriall water*, which is the true tincture. For this separation of the pure from the impure, is not done with *hands*, but *nature* her selfe alone, by working it circularly to perfection, bringeth it to passe. It appeareth then that this composition is not a manuall worke, but onely a change of the natures, because *nature* dissolves and conjoynes it selfe, it sublimes and lifts up it selfe, and having separated the *fœces*,[72] it groweth *white*: and in such a sublimation the parts are alwayes joyned together, more subtile, more pure and essentiall, because that when the fiery nature lifteth up the subtile parts, it lifteth up alwayes the more pure, and by consequent leaveth the grosser in the bottome. And therefore it behooveth by an indifferent fire, to sublime in a continuall vapour, that the *Stone* may bee inspired in the *ayre*, and live. For the nature of all things takes life of the inspiration of *ayre*, and so also all our *Maistery* consists in vapour, and in the sublimation of *water*. And therefore our *brasse* or *Leton* must by degrees of fire bee lifted up, and freely without violence, of himselfe, ascend on high, wherefore unlesse the *Body* bee by *fire* and *water* dissolved, attenuated, and subtilized, untill it ascend as a *spirit*, or climbe like *argent vive*, or as the *white soule* separated from the *Body*, and carried in the sublimation of the Spirits, there is nothing at all done in this *Arte*: But when it ascends on high, it is borne in the *ayre*, and changed in the *ayre*, and is made *life* with *life*, being altogether spirituall and incorruptible: And so in such a regiment the *Body* is made a *spirit* of a subtile nature, and the *spirit* is incorporated with the *Body*, and is made one with it, and in such a sublimation, conjunction, and elevation, all things are made *white*.

And therefore this *Phylosophicall* and naturall sublimation is necessary, for that it maketh peace betweene the *body* and the *spirit*, which is unpossible otherwise to be done, otherwise then by this separation of the parts: wherefore it behoveth to sublime them both, to the end, that in the troubles of this stormy Sea,[73] the *pure* may *ascend*, and the *impure* and earthly may *descend*: And for this cause it must be boyled continually, that it may be brought to a subtile nature, and that the *body* may assume and draw to it selfe the *white Mercuriall soule*, which it naturally retaines, and suffereth it not to be separated from it, because it is like unto it, in the neerenesse of the first, pure, and simple nature. From hence it appeares, that this separation must be made by decotion, untill there remaine no more of the *fat* of the *soule*,[74] which is not lifted up, and exalted into the upper part, for so they shall be both reduced unto a simple *equality*, and unto a simple *whitenesse*. *The vulture therefore flying in the ayre, and the Toade going upon the Earth,*[75] *is our Maistery*: And therefore when thou shalt gently, and with great discretion, separate the *Earth* from the *water*, that is, from the *fire*, and the *subtile* from the *thicke*, then that which is *pure*, will *ascend* from Earth into Heaven, and that which is *impure*, will goe downe to the *Earth*, and the more subtile part will in the *upper* place take the nature of a *spirit*, and in the *lower* place the nature of an *Earthly Body*; wherefore let the *white nature* with the more subtile part of the *Body*, be by this operation lifted up, leaving the *fœces*, which is done in a short time: for the *soule* is aided by her associate and *fellow*, and perfected by it. *My Mother* (saith the *Body*) *hath begotten mee, and by me shee her selfe is begotten; and after shee hath taken her flight, (or I have taken from her her flying) shee after the best manner shee can, becomes a pious Mother, nourishing and cherishing the sonne whom shee hath begotten, untill he come to perfect state.* Heare this secret: Keepe the *Body* in this our *Mercuriall water*, untill it ascend on high with the *white soule*, and the Earthly descend to the bottome, which is called, *the Earth that remaines*: then shalt thou see the water coagulate it selfe with its *body*, and shalt bee assured that the *Science* is true, because the *Body* coagulateth his moisture into the drinesse, as the rennet of a *Lambe* coagulateth milke into *Cheese*.[76] In the same fashion the *spirit* will pierce the *body*, and there will be a perfect mixture made by the least parts, and the *Body* will draw unto himself his *moisture*, that is to say, his *white soule*, even as the *Load-stone* draweth the *Iron*, because of the likenesse and neerenesse of his nature, and his greedinesse, and then the one will hold the other, and this is our sublimation and coagulation, which retaineth every thing *volatile*, and maketh that it can flye no more. Therefore this composition is not a manuall operation, but (as I said) a changing of natures, and a wonderfull connexion of their *cold* with *hot*, and their *moist* with *dry*: for the *hot* is mixed with *cold*, and the *dry* with *moist*, and so by this meanes is made the mixture and conjunction of the *body* with the *spirit*, which is called the changing of *contrary natures*; because that in such a solution and sublimation, the *spirit* is turned into a *body*, and the *body* into a *spirit*; so that the natures being mingled together, and reduced into one, doe change one another, in as much as the *body* makes the *spirit* a *body*, and the *spirit* turnes the *body* into a teyned and *white spirit*.

And therefore (this is the last time that I will tell thee) boyle it in our *white water*, that is, in *Mercury*, untill it bee dissolved into *blacknesse*, and then by continuall decoction, it will bee deprived of his *blackensse,* and the *body* so dissolved, wil at length arise with the *white soule*, and then one will bee mingled with the other, and they will embrace one another, so that they shall no more be divided asunder, and then the *spirit* is united to the *body* with a reall accord, and are made one *permanent* thing; and this is the *solution* of the *body*, and the *Coagulation* of the *spirit*, which have one and the selfe same operation.

Hee therefore that knoweth how to *mary*, to *make with childe*, to *mortifie*, to *putrifie*, to *engender*, to *quicken* the *species*, to bring in the *white light*, and to *clense* the *vulture*[77] from his *blacknesse* and *darknesse*, untill he be purged by *fire*, coloured and purified from all his spots, shall bee the owner of so great dignity, that *Kings* shall reverence him, and doe him honour.

Wherefore let our *body* abide in the *water*, untill such time as it be loosed into a new *powder* in the bottome of the *vessell* and of the *water*, which is called the *blacke ashes*, and this is the corruption of the *body*, which is by wise men called *Saturne, Leton,* or *Brasse*, the *Phylosophers Lead*, and the *discontinued powder*.[78] And in this putrifaction and resolution of the *Body*, there appeare *three* signes, to wit, the *blacke colour* the *discontinuity* of the parts, and a *stinking smell*, which is likened to the smel of *sepulchres* or graves.[79] This ashes then is that of which the *Phylosophers* have said so much; which remained in the lower part of the vessell, *which wee ought not to despise*, for in it is the *Diademe* of our *King*, and the *Argent vive*, blacke and uncleane, from whence the *blackensse* must be purged by continuall decoction in *our water*, untill it be lifted up in a *white colour*, which is called the *Goose*, and the *Poulet* of *Hermogenes*.[80] He therefore that maketh the *red Earth blacke*, and then *white*, hath the *Maistery*, as also hee that *killeth* the *living*, and *quickeneth* the *dead*: therefore make the *blacke white*, and the *white red*, that thou mayest make the worke perfect; and when thou seest the true *whitenesse* appeare, which shineth like a naked *Sword*, know that in that *whitenesse*, is *rednesse* hidden; and then thou must not take out of the vessell that *whitenesse*, but onely boyle it, to the end, that with drinesse and heate, there may come upon it a *Citrine* colour, and in the end, a most shining and sparkling *red*; which when thou seest, with great feare and trembling, praise the most good, and most great *God*, which giveth wisedome, and by consequence, riches unto whom he pleaseth; and according to the iniquity of the *Persons*, taketh them away againe, and depriveth them of them for ever plunging them in the servitude and slavery of their enemies. To him be praise and glory for ever and ever. *Amen.*

FINIS.

THE EPISTLE
of JOHN PONTANUS,

(*mentioned in the Preface to the Reader of*
ARTEPHIUS his secret Booke)
wherein he beareth witnesse of the BOOKE:
Translated out of the Latine Copy:
Extant in the third Volume of
Theatrum Chymicum, at the 775. Page.

I John Pontanus,[1] *have traveiled thorow many Countries, that I might know some certainty of the Philosophers Stone; and going thorow as it were all the world, I found many false deceivers, but no true Philosophers, yet continually studying, and making many doubts, at the length I found the trueth: But when I knew the matter in generall, I yet erred two hundred times, before I could attaine to the true matter, with the operation and practise thereof. First I begunne to worke with the matter, by putrefacton nine moneths together, and I found nothing: Then I put it into* Balneum Mariæ *for a certaine time, and therein I likewise erred: Afterwards I put it in the fire of calcination for three moneths space, and I wrought amisse. I tryed all kinds of distillations and sublimations, (as the Philosophers,* Giber,[2] Archelaus,[3] *and all the rest, either say or seeme to say) and I found nothing. In summe, I assayed to perfect the* Subject *of the whole Art of Alchimy, by all meanes possible*

81

to be devised, as by Dung, Bathes, Ashes, and other fires of divers
kinds, which yet are all found in the Philosophers Bookes, but I
found no good in them. Wherefore I studied three whole yeeres in
the Bookes of the Philosophers, especially in Hermes alone, whose
briefer words doe comprehend the whole Stone, though hee speake
obscurely of the superior, and inferiour, (or that which is above,
and that which is below)⁴ of heaven & earth. Therefore our Instrument which bringeth the matter into being in the beginning,
second, and third worke, is not the fire of a Bath, nor of Dung,
nor of Ashes, nor of the other fires which the Philosophers have
put in their Bookes: What fire is it then which perfects the whole
worke from the beginning to the ending? Surely the Philosophers
have concealed it: But I being mooved with pitie, will declare it
unto you, together with the complement of the whole worke. The
Philosophers Stone therefore is one, but it hath many names, and
before thou know it, it will be very difficult; for it is watery, aiery,
fiery, earthy, flegmaticke, cholericke and melancholy; for it is sulphurous, and it is likewise Argent vive, and it hath many super-
fluities, which by the living God are turned into the true essence,
our fire being the meanes: And hee that separates any thing from
the subject, thinking it to bee necessary, hee truely knoweth nothing at all in Philosophy; for that which is superfluous, uncleane,
filthy, fœculent, and in summe, the whole substance of the Subject, is perfected into a fixt spirituall body, by the meanes of our
fire. And this the wise men never revealed, and therefore few doe
come unto the Arte, thinking that there is some such superfluous
and uncleane thing. Now wee must seeke out the properties of our
fire, and whether it agree to our matter, after the manner that I
have sayd, to wit, that it may bee transmuted, when as that fire
doth not burne the matter, it separateth nothing from the matter,
it divideth not the pure parts from the impure, as all the Philosophers say, but it turneth the whole Subject into puritie. It doeth
not sublime, as Geber maketh his sublimations; Arnold⁵ likewise
and others speaking of sublimations and distillations, to bee done
in a short time. It is minerall, equall, continuall, it vapours not,
except it bee too much stirred up: it partaketh of Sulphur, it is
taken from else-where then from the matter; it pulleth downe all

things, it dissolveth and congealeth, likewise it both congeales and calcines, and it is artificiall to finde out, and is a compendious and neere way, without any cost, at least with small cost: and that fire is it, with a meane firing, for with a soft fire all the whole worke is perfected, and it performeth withall, all the due sublimations. They that should reade Geber, *and all the other* Philosophers, *though they should live an hundred thousand yeeres, could not comprehend it, because that fire is found by deepe and profound Meditation onely, and then it may be gathered out of Bookes, and not before. And therefore the errour of this Arte is, not to finde the fire, which turnes the whole matter into the true Stone of the* Philosophers. *And therefore studie upon it, for if I had found that first, I had never erred two hundred times, in my practise upon the matter: wherefore I doe not mervaile, if so many and great men have not attained unto the worke. They doe erre, they have erred, they will erre, because the* Philosophers *have not put the proper* Agent, *save onely one, which is named* Artephius, *but hee speakes for himselfe, or by himselfe; And unless I had read* Artephius, *and felt him speake, I had never come to the complement of the work. But the* practique *is this. Let it bee taken; and ground with a physicall contrition, as diligently as may bee, and let it bee set upon the fire, and let the proportion of the fire bee knowne, to wit, that it onely stirre up the matter, and in a short time, that fire, without any other laying on of hands, will accomplish the whole worke, because it will putrifie, corrupt, ingender; and perfect, and make to appeare the three principall colours, blacke, white, and red. And by the meanes of our fire the Medicine will bee multiplied, if it be joyned with the crude matter, not onely in quantitie, but also in vertue. With all thy strength therefore, search out this fire, and thou shalt attaine thy wish, because it doeth the whole worke, and is the Key of the* Philosophers, *which they never revealed: But if thou muse well and profoundly upon those things that have beene spoken concerning the properties of the fire, thou mayest know it; otherwise not. I beeing mooved with pitie, have written these things, but that I may satisfie thee fully, this fire is not transmuted with the matter, because (as I said above) it is not of the matter. These things therefore I thought fit to say, and to warne the prudent, that they spend not their moneys unprofitably, but know what they ought to looke after. For by this meanes they may come to the truth of the Arte, and not otherwise.* Farewell.

FINIS.

Commentary

Epistle Dedicatory

[1] *under-Moone.* Occult or mysterious—"sublunary" in the sense that all substance beneath the moon's sphere, in Ptolemaic cosmology, is mutable and perishable, as opposed to the permanence that exists beyond it.

[2] The history of alchemy notes several women philosophers. The most famous is the legendary Maria (Miriam) Prophetissa, also called the Jewess, sister of Moses or the Copt, and described by Zozimos of Panapolis as a Graeco-Egyptian alchemist of the third century. Maria's works do not survive, though a treatise is ascribed to her under the title "Mariae Prophetissae practica" (in *Artis auriferai* I: 343–8). She is credited with the invention of the water-bath (*bain-marie*) and the *kerotakis*. A modern "portrait" of Maria Prophetissa appears in the frontispiece of M. Maier, *Symbola aureae mensae*, and she is also shown supervising the process of *coniunctio* further on in the treatise. Other female alchemical philosophers mentioned in Mylius, *Opus medico-chemicum* are "Cleopatra, Queen of Egypt," "Medera, the woman Alchemist," "Thaphuntia, the woman Philosopher," and "Euthica, the Arabic woman Philosopher."

[3] This passage from Zecharia prefigures the coming of Christ the judge in the book of Revelations.

[4] The 1612 French edition of Flamel's *Exposition of the Hieroglyphicall Figures* attributes this saying to Virgil.

Booke of the Hieroglyphicall Figures

[1] *which lifteth the humble from the base dust.* A reference to God's creation of Adam from earth (Genesis 2:7) and to the resurrection of the flesh from dust (Ecclesiastics 12:7)—both apt synonyms for the alchemical work.

[2] *worldly Sphaeres (or circles).* The twelve elemental, planetary and zodiac spheres, perhaps alluding to the enduring felicity in heaven as opposed to inferior "earthly" pleasures which the pious put behind them.

[3] *one thousand three hundred fourscore and nineteene.* 1399.

[4] *dressing accounts.* Drawing up accounts.

[5] "10 s. (hilings) English," 1692.

[6] *Rindes.* "*escorces,*" 1659; *OED* defines "rindes" as bark of a tree or plant, here perhaps a reference to papyrus.

[7] *Gaule.* "French," 1692.

[8] *point of iron.* Ordinarily a burin used to scratch lines into a metal plate for subsequent engraving of multiple copies. However, in this case, where one-of-a-kind drawings are described, the "point of iron" perhaps refers to silver point, a drawing implement that achieved its effect by means of silver oxide.

[9] For analysis of the first seventh, second seventh and last seventh leaves of Abraham's book, see intro., pp. xxiii–xxv.

[10] *Maranatha.* "If any man love not the Lord Jesus Christ let him be Anathema Maran-atha" (Corinthians 1:22).

[11] See intro., p. xxii on the persecution of Jews in France.

[12] *by the sides.* In the margins.

[13] *first Agent.* The *prima materia* or "first matter."

[14] *Cabala.* The name given in post-Biblical Hebrew to the oral tradition handed down from Moses to the Rabbis of the Mishnah and the Talmud (*OED*). "That more secret and true exposition, forbidden to be set down in writing, which Moses divinely received upon the Mount, together with that other law which was afterwards given in writing to the people of Israel." Indeed, Glanatinus asserts that by means of this Cabala "the ancient Rabbins came to the knowledge of the Trinity, and of Christ the Son of God" (Rulandus, 76–77).

[15] *enlightened.* Illuminated. "*enluminees*," 1659.

[16] *Caducean rodde.* "The Caduceus is composed of three parts—of the golden stem, or rod, surmounted by an iron apple, and of two serpents which seem to be on the point of devouring it. One of these serpents represents the volatile portion of the matter of the Philosophers, the other signifies the fixed part, and these strive with one another in the vessel. They are united, equilibrated, and restrained in the poise of fixation by the philosophical gold, typified in the stem or rod, and thus they are inseparably united in one body" (Rulandus, 344).

Mercurius is illustrated juxtaposed with the caduceus in Catari, *Le imagini*, 400 and Basilius Valentinus, *Duodecim claves*, in Maier, *Tripus aureus*, 396.

[17] See intro., p. xxv. An image of dismemberment as a metaphor for the "fixing" of mercury appears in Trismosin, *Aureum Vellus*. Manuscript prototypes appear in a late sixteenth-century manuscript of Trismosin's *Splendour Solis* (1582), London, British Museum, MS. Harley 3469 and in the Pseudo Thomas Aquinas, *Aurora consurgens*, Zurich, Zentralbibliothek, Cod. Rhen. 172, 27. See Barbara Obrist, *Les Débuts de l'imagerie alchimique* (Paris, 1982), 275–84.

[18] For explanations of the images on the verso of the fourth and the recto of the fifth leaves, see intro., p. xxv–xxvii.

[19] *Fauchion.* Or Falchion—"A broad sword more or less curved with the edge on the convex side" (*OED*). "*coutelas*," 1659.

[20] See intro., p. xxix for discussion of this emblem. "Alchemists understand by the term Bathing the Coction of the Matter, and its Cir-

culation in the Philosophical Egg . . . Calcination, Tinging, Washing, Whitening and Bathing signify one operation, and that all these words signify the Coction of the Matter until it has attained its perfection . . . Bath is a matter reduced into a liquid or aqueous form; thus when it is required to perform projection on a metal, that metal must be melted, and this is called the Bath, or reduction into the mercurial form, where the King and Queen come to Bathe—that is, the Sun and Moon—because it is liquid water" (Rulandus, 430).

Related images appear in many illustrated alchemical treatises, most notably: *Rosarium Philosophorum* in *Artis Auriferae*, II, 262 and Maier, *Atalanta*, Emblem XXXIV. Manuscript illustrations of the sun and moon bathing appear in the *Rosarium philosophorum*, St. Gallen, Stadtbibliothek Vadiana, MS. 394a, fol. 34; *Adamas colluctancium aquiliarum*, Biblioteca Apostolica Vaticana, MS. Pal. Lat. 412, fol. 57; and in the various versions of the *Ripley Scrowle* (Oxford, Bodleian Library, MS. 1588 and London, British Library, MSS. Add. 5025 and Add. 32621).

[21] *he who . . . wished that all the men of the World had but one head etc.* Probably refers to Caligula's statement "Utinam populus Romanus unam cervicem haberet" (Seutonius, *Gaius Ceasar*, 30).

[22] *not knowing with what matter I should beginne.* Not knowing of the *prima materia*.

[23] *gravings.* Not "engravings," but drawings.

[24] *Licentiate in Physick.* Doctor of medicine.

[25] *Argent vive.* The chemical term for mercury (quicksilver) (Rulandus, 46).

[26] *fixe.* To "fix" quicksilver is to solidify it.

[27] *bloud of young infants.* Not a reference to infanticide, but to *aqua permanens*, made by the philosophical solution of two perfect metallic bodies, gold and silver, dissolved in water and united (Rulandus, 34). Alchemical imagery of *aqua permanens* is often based upon the blood symbolism and allegories of the Church, in this case, the Massacre of the Innocents.

[28] *an herb.* The "Herb" described as appearing on the verso of the fourth leaf of Abraham's book is a flower with blue, white and red flowers and gold leaves. This description distinguishes the three major colors of alchemy—blue for putrefaction, white for cleansing and red the color

of the lapis—and their unification into a precious (gold) substance finer than its components. Illustrations of flowers as visual metaphors for the alchemical process appear in Reusner, *Pandora*, 257; Mylius, *Anatomiae*, 20 and idem *Opus medico-chymicum*, seal of "Jesid Constantinopolitanus"; Boschius, *Symbolographia*, LCCXXIII; Fludd, *Summum bonum*, frontispiece. An early manuscript prototype is the alchemical flower in the *Ripley Scrowle*, Oxford, Bodleian Library, MS. 1588 and fol. 13 of Johannes Andrae, London, British Museum, MS. Sloane 2560.

[29] *Serpents.* Symbols of the corrosive by-products of putrefaction, perhaps an acid or alkali. Allegorically, this passage could also describe philosophical Mercurius existing as the *lapis* in putrefaction, where, after mediating the opposites (gold and silver) it takes on the guise of a poisonous dragon or serpent before its decoction and reduction. Images of dragons are ubiquitous in the alchemical illustrative tradition. See: *Aurora consurgens*, Zurich, Zentralbibliothek, Cod. Rhen. 172, fol. 36; Nazari, *Della tramutatione*, Fig. 16; Maier, *Atalanta*, emblems XXV and XXIX; and Mylius, *Anatomia*.

[30] *St. James of Gallicia.* St. James of Compostella, patron saint of pilgrims.

[31] *Iris, or the Rainebow.* The many colors of the *lapis* were compared poetically to a rainbow in "Gloria mundi," *Musaeum hermeticum*, 251, (Waite, I:202): "An earthly manifestation of the quintessence you may behold in the colours of the rainbow, when the rays of the sun shine through the rain." In the Classical tradition, Iris is the goddess of the rainbow and messenger of the greater gods presumably because the rainbow seems to touch both sky and earth (*OCD*). Alchemically, Olympiodorus linked Iris and rainbows, saying " . . . the pupil of the eye and Iris (rainbow) in the sky" (Berthelot, II, iv:38). The many colors of the *lapis* which were said to appear near the completion of the alchemical work were also called the "peacock's tail" (see note 169). Khunrath refers to all the colors as "Iris, the messenger of God, and the peacock's tail" (*Amphitheatrum sapientiae*, 202). According to Jung (*Mysterium coniunctionis*, 290) Iris, the messenger of the gods, heralds the *lapis*, which follows the alchemical rainbow just as the rainbow was a sign of God's covenant with Noah after the flood (Genesis 10:12).

[32] *Montjoy.* "mont-joie," 1659. A Cairn, or hill serving as an observation post. See intro., p. xxi.

[33] *Leon.* Perhaps Laon.

³⁴ *Boloyn.* "Boulogne," 1659.

³⁵ On Canches, see intro., p. xxii.

³⁶ *Sanson.* Perhaps Saxony.

³⁷ See intro., p. xxii on the numerical symbolism of Canches' seven-day sickness.

³⁸ *projection.* "Exaltation of a substance by a Projecting Medicine — which is projected over the matter to be transformed . . . performed by a violent interpenetration, which transforms at the moment of ingression. Moreover, the Medicine is not called a Ferment, but a Tincture" (Rulandus, 263).

³⁹ *17 of January about noone.* See intro., p. xxii on the meaning of the date of this projection.

⁴⁰ *red stone.* Sulphur (Rulandus, 306).

⁴¹ *to conquere this rich golden Fleece.* The alchemical work is here compared to the dangerous voyage of Jason and the argonauts and their triumphal discovery of the golden fleece (the *lapis*). As an alchemical allegory of the danger and uncertainty of the work, the event is pictured in Maier, *Atalanta*, emblem XLIX and described in Maier, *Symbola aureae mensae*, 35. A parallel exists between Jason's adventures and Flamel's and Canches' dangerous sea voyage back to France.

⁴² *amongst the bones of the dead.* An allusion to the charnel house setting of the sculpted tympanum.

⁴³ *Hermes.* The name Hermes Trismegistus is a translation of the Egyptian "Thoth the very great." Hermes is the reputed author of the philosophical-religious treatises known collectively as *Hermetica*, among which the *Tabula smaragdina* (*Emerald Tablet*—printed in 1541) is the best known alchemical work. These writings are accepted as being Greek, rather than ancient Egyptian, their attribution to the Egyptian god Thoth being a renaissance invention (*OCD*).

⁴⁴ *a man all blacke.* Alchemical iconography often personified the blackness of Saturnine *putrefactio* as a Negro, Moor, or Ethiopian. The black man also symbolizes the base *prima materia* upon which the entire operation depended. Trismosin's *Aureum vellus*, 181-182, describes the materials in *nigredo* as "a man, black like a Negro." As the Ethiopian in the *Aurora consurgens*, he cries out, in words reminiscent of the biblical "Song of Songs," "Be turned to me with all your heart and do not cast

me aside because I am black and swarthy" (Aquinas (pseud.) *Aurora consurgens*, I:xii). The charred ingredients in *putrefactio* assumed the persona of an Ethiopian in Melchior Cibinensis' alchemical mass (*Symbolum* in Maier, *Symbola aureae mensae*), where the death, purification, and regeneration of the materials are described as follows: "Then will appear in the bottom of the vessel the mighty Ethiopian, burned, calcined, discoloured . . . He asks to be buried, to be sprinkled with his own moisture and slowly calcined till he shall arise in glowing form from the fierce fire . . . Behold a wondrous restoration and renewal of the Ethiopian!" Maier, *Atalanta*, 58, calls the "sacred Ethiopian" of alchemy "the touchstone of truth." Jung (*Mysterium coniunctionis*, 401–2) links the origin of this metaphor to a treatise attributed to Albertus Magnus, "Super arborem Aristotelis" (*Theatrum chemicum*, II, 526). The passage runs ". . . until the black head bearing the resemblance of the Ethiopian is well washed and begins to turn white . . . " Illustrations of the Ethiopian occur in Trismosin's *Splendor solis*, British Museum, MS. 1582, and in Mylius, *Opus medico-chymicum*, seal of "Euthyices Philosophus."

[45] *vessell of Phylosophy*. An Athanor, or digesting furnace in which constant heat is maintained by means of a tower which provided a self-feeding supply of charcoal (*OED*). Rulandus, 52–3, gives a lengthy description: "The fire does not touch the base, and the required heat is suitably and uniformly imparted . . . A circular wall is erected of the height of one foot. On either side of this wall a vacant space, with a small door, is left. This for the removal of the ashes. Above this structure there is placed a small iron gridiron, and above the said gridiron we erect another small door, which is broader at the bottom than at the top, and is an aperture through which the coals can be stirred with a poker. When this turret has been set in an upright position in the manner described, and has been filled with coals to the top we cover it with a covering of clay. But at the same time, in the hindmost part of the wall, and in that portion of it which is nearest to the gridiron, we leave a small hole open, through which the heat may be able to approach the Athanor, and we stop this aperture with a spatula (a long instrument for stirring), or with an iron bar (some term it a register), which can be raised and lowered. We make also at the top of the turret, of the breadth of five inches, beneath the cover, a small aperture, through which the index finger shall just be able to pass, by which the fire may draw the air, as if a fuel, to itself. Over against the turret constructed in the manner described, there is set another oven, which is the Athanor itself.

After the same manner, a circular wall, one foot and a half in height, is constructed, which completely fits with its sides the posterior opening of the turret. On this wall we erect an oven, leaving on the top of the furnace a small aperture, like an imperial thaler, whereby the heat in this part, to some extent pressing upon the furnace, can pass upward to the next nearest furnace. Then we again build an eighteen-inch wall by the place where we commenced the furnace; we cover the same with a lid, again leaving a small aperture at the top, as in the case of the lower one. However, it is necessary that in one side of this part there should be left a clear space where the matter can be put in and taken out. For in this middle part is the workshop where the matter is prepared in its proper vessel, placed over a tripod. In order to fill up the clear space, and close it up lest any air should be produced, a well-fitting lid is made to cover it. Finally, with a third lid, we cover the whole of this second furnace, leaving, however, at the base four air-holes, which also have covers, whereby the heat may be increased or diminished."

Illustrations of athanors appear in Gesner, *New Book of the Distillatyon of Waters*; Libavius, *Alchymia*, title page; Michelspacher, *Cabala*, title page; Valentinus, *Duodecim clavibus* in Maier, *Tripus aureus*, fig. 13 and several emblems in Mylius, *Philosophia reformata*.

[46] *Penner.* A pen-case (*OED*).

[47] *Scutchions.* Escutcheon. *armoires*, 1659. "The Shield or shield-shaped surface on which a coat of arms is depicted" (*OED*).

[48] *Innocents were killed by command of King Herod.* Matthew 2:1–22; Luke 1:5.

[49] *sinnes which naturally are enterchayned, etc.* A reference to original sin, which attends all people at birth, and those sins acquired during the course of a lifetime of temptation and transgression.

[50] *triple Vessell.* The symbolism of the number three is very important in alchemical imagery. Like the Christian Trinity of Father, Son and Holy Spirit, the final product of the alchemical work was ". . . triple in name but one in essence" (London, Wellcome Institute for the History of Medicine Library, Wellcome MS. 2456, fol. 332). Among the various trinitarian parallels are: animal, mineral and vegetable; sun, moon and Mercurius; body, soul and spirit; magnesia, sulphur and mercury; or sulphur, mercury and salt. Illustrations of the alchemical "three in one" are legion. Manuscript versions include several emblematic illustrations in the *Book of the Holy Trinity*, St. Gallen, Kantonsbibliothek Vadiana, MS.

428, especially fol. 22 verso; *Das Buch der heiligen Dreifaltigkeit*, Munich, Staatbibliothek, MS. Germ. 598; Trinity College, Cambridge, MS. O. 8. 24, fol. 4 (See Obrist, 261–275). Printed illustrations include the *Rosarium philosophorum*; Nazari, *Della tramutatione*, fig. 16; Lambsprinck, *De lapide philosophica*, Emblems 11 and 15; and Mylius, *Philosophia reformata*, emblem 7.

[51] *Philosophicall Egge.* Continuity between the organic and inorganic worlds was a basic alchemical concept, and changes in substances were likened to changes in a growing chick embryo. The vessel in which the change took place was often termed the "philosophical egg," so named because of its shape. According to Ripley, *Opera omnia chemica*, 30, "In uno vitro debent omnia fieri, quod sit forma ovi" (Everything must be done in one glass, which must be egg-shaped). The materials were cooked and heated in this vessel in imitation of the internal heat thought to generate life in a hen's egg. Illustrations of actual egg-shaped apparatus are common. See especially *Glorieuse Marguerite*, Bibliothèque Nationale, Paris, MS Fr. 1089, fol. 1 verso and many illustrations in an anonymous fifteenth-century technical treatise, British Museum, London, MS Harley 2407, fols. 108–10. An interesting allegorical image of an egg that houses the opposites personified as eagles wearing the crowns of a pope and an emperor appears in a manuscript treatise ascribed to Wynandi de Stega, *Adamas colluctancium aquilarum*, Biblioteca Apostolica Vaticana, MS Pal. Lat. 412, fol. 85 verso. See also Maier, *Atalanta*, Emblem VIII.

[52] *Scumme of the Red Sea . . . fat of the Mercurial wind.* The solid by-products of sulfur and mercury in a gaseous state.

[53] *Three vessels.* "the furnace, the sand vessel, and the philosophical egg," 1692.

[54] *Athanor.* see note 45.

[55] *Balneum Mariae.* The "Mary's bath" was supposedly invented by the legendary Maria the Jewess. It functioned as a gentle water bath and is described by Rulandus, p. 69, as "The dissolution of a substance in a suitable vessel of warm water, after which it is placed in the copper vessel belonging to it." Illustrations of the apparatus appear in Libavius, p. 80. See also the reconstruction in F. Sherwood Taylor, *The Alchemists* (New York, 1949).

[56] *greene Lyon.* Yet another incarnation of Mercurius, personified earlier in the text as *aqua permanens* and the "Serpent that devours, stiffens, and mortifies his own tail" (Maier, *Symbola aureae mensae*, 427).

It is green vitriol, a strong caustic dissolving agent and the violent allegorical persona of the medium of conjunction of alchemical opposites. De Jong traces the source of the image to Morienus, *De transmutatione metallorum* (*Artis auriferae*, II:55): "Ecce iam nomina specierum tibi exprimo quarum tamen tres ad totum magisterium cibi sufficient: id est fumus albus, et leo viridis et aqua foetida, ("Behold! I am already telling you the names of the kinds, three of which may be enough for you for the whole masterpiece, they are white smoke, the green lion and stinking water.")

The green lion appears in both manuscript and printed versions of the *Rosarium philosophorum* (*Artis auriferae*, II:236) and is described and illustrated in Maier, *Atalanta*, emblems XXXVII and XXVIII (See de Jong, 247–51). See especially the color illustrations in the sixteenth-century manuscript *Rosarium*, St. Gallen, Stadtbibliothek Vadiana, MS 394 a, fol. 97 (later reproduced in J. D. Mylius, *Philosophia reformata*, emblem 18) and in the *Ripley Scrowle*, London, British library, MS Add. 32621.

[57] *phioll*. "A glass vessel having a globular body and a long slender neck or funnel" (Rulandus, 249).

[58] *Bolts-head. Curcubite*, 1659; "A vessel named for its resemblance to a common gourd (cucurbit), shaped for the most part like an inverted cone. One form is globe-shaped at the bottom; another is flat" (Rulandus, 120).

[59] *Summarie or Abridgement of Philosophy*. The earliest edition of Flamel's *Summarium philosophicum* is *Le Sommaire philosophique*, in *Trois anciens traites en rytmes françaises*, ed. Jacques Gohorry, Paris, 1561. Early seventeenth-century editions are in *Vier nuetzliche Chymische Tractat*, 1612, sig. Fvj verso and *Wasserstein der Weysen*, 1619, 214.

[60] *Poulet*. Chicken. Mylius, *Philosophia reformata*, 117, refers to the chicken as the bird of Hermes, another synonym for the mercurial serpent and lion. The *Turba of the Philosophers* describes the chicken as a metaphor for the entire alchemical process, and instructs the practitioner to "Put our material in an egg. There will happen a chicken with a red crest, white body, and black feet" (*La Tourbe*, 34).

[61] *belly*. The "belly of the horse" is a place of decoction and digestion (John of Rupecissa *La Vertue et propriété de la quinte essence*, 94). An allegorical illustration of the concept appears in the anonymous *Traité d'alchimie*, Paris, Bibliothèque Arsenal, MS 6577, fol. 8 verso.

[62] *wombe.* Womb imagery here refers to the belief that alchemical materials "grow" within flasks and furnaces until they reach a state of maturity, the *lapis* or, in the case of metals, gold. For an explanation in Paracelsan terms, see Kelly, *Basil Valentine*, xxvi.

[63] *yong King.* Immature *lapis.*

[64] *Calid the Persian.* Seventeenth-century tradition claimed Kalid (Kalid ben Jazichi) to have been a Jew. His works were believed to have been translated from Hebrew into Arabic and then into Latin. No manuscript sources of his works exist dating before the sixteenth century, however, when works attributed to him began to be printed. The first edition of Kalid's *Liber Secretorum Alchemiae* appeared in the collection *Alchemia* (1541), 338. It was subsequently reprinted in 1545, 1561 and 1610 (Ferguson, I:448-449). The treatise was translated into French in 1557 and into English in *The Mirror of Alchimy, Composed by the thrice-famous and learned Fryer, Roger Bachon* (London, 1597). See Linden, 28-53.

[65] *Morien.* Morienus was a legendary Arabic philosopher whose *Liber de compositione alchemiae* was the first alchemical treatise translated from Arabic into Latin (by Robert of Chester in 1144). His writings may have come to the West about the time of the Crusades. They were also translated from Arabic into Latin in 1182 by Robertus Castrensis (Ferguson, II:108-9).

[66] *flowres.* Vapours. Alchemically, the flower (*flos*) is the volatile, spiritual substance of a thing (Rulandus, 147).

[67] *according to the Winter.* A very low heat.

[68] *Aries to . . . Cancer.* The Spring months, late March to mid-June.

[69] *cold flegme.* In humoral theory, phlegm is the cold, wet body fluid linked with the season of winter and the moon.

[70] *bath.* See intro., p. xxix on alchemical opposites in a bath.

[71] *dry Choller.* In humoral theory, cholor is the hot, dry body fluid associated with yellow bile and fire.

[72] *Rasis.* Abu Bekr Muhammed Ben Zakeriyah al-Rasi, called Rasis (also Razis, Rhasis, or Rhazes) was born in Iraq c. 850-60 and died c. 923. He was known during his life as a great physician and philosopher and the influence of his medical knowledge passed into the Renaissance.

Rhasis is reputed to have written several alchemical works: *Duodecim libri de arte chemica; Arcanorum liber; Perfectionis liber; Liber lapidis minor;* and *Confirmatio artis chimiae.* His treatises were printed in the mid-sixteenth century (Ferguson, II:262–3).

[73] *golden Bird.* Perhaps the alchemical phoenix, metaphor for the lapis. According to Ovid (*Metamorphoses,* XV, 392–407), the phoenix visits Heliopolis in Egypt from his home in Arabia every five hundred years, at which time he builds himself a pyre and immolates himself. He then arises out of the ashes as a symbol of the rebirth of the *lapis* from putrefaction. The phoenix is illustrated in Trismosin, *La Toyson d'or,* Fig. 13; Boschius, *Symbolographia,* symbol DCVI1792; the title page of Maier, *Jocus severus*; the title page of *Musaeum hermeticum,* 1625; and the title page of Campy, *L'Ouverture.*

[74] *Diomedes.* A doctor mentioned by Galen as suggesting a collyrium to him (Kelly, 170).

[75] *to Cancer . . . toward Libra.* Mid-June toward mid-September.

[76] *Libra towards Capricorne.* Mid-September towards early December.

[77] *two Dragons.* Personifications of the alchemical opposites, here distinguished as mercury and sulphur. When they are joined into the *uroborus,* the dragons also become a metaphor for circular distillation. For illustrations of the *uroborus,* see Eleazar, *Uraltes chymisches Werk,* Part II, no. 4 facing page 8; also the *Ripley Scrowle,* London, British Museum, MS. Add. 5025. The concept is sometimes pictured as two lions, one winged (Maier, *Atalanta,* Emblem XVI; Lambsprinck, *De lapide philosophica,* emblem IV; Mylius, *Philosophia reformata,* 1622, emblem 23; Valentinus, "XI Clavis," *Chymische Schrifften,* 68), a wolf and a dog (Maier, *Atalanta,* Emblem XLVII) or as a winged and wingless bird (Maier, *Atalanta,* Emblem VII).

[78] *Argent Vive.* See note 25.

[79] *head biting the tayle.* For manuscript sources illustrating the single *uroborus,* see a fifteenth-century copy of Synosius by Theodoros Pelecanus, Paris, Bibliothèque Nationale, MS. grec. 2327, fol. 297 and Pseudo Thomas Aquinas, *De alchimia,* Leiden, Bibliotheek der Rijksuniversiteit, Cod. Voss. Chym. 29, fol. 94a. Printed images include Mylius, *Philosophia reformata,* emblem 3; Eleazar, *Uraltes chymisches Werk,* Part II, no. 3 facing p. 8; and Reusner, *Pandora,* 257.

[80] *Hesperides.* The gardens of Hesperides refers to Hercules' task of plucking the golden apples from the dragon-guarded tree at the world's end (*OCD*). Hercules as a model of the alchemist in the Garden of the Hesperides appears on the title page of Maier, *Arcana arcanissima.* The apple tree in the Hesperides garden also appears in Boschius, *Symbolographia,* LVII.

[81] *Jason.* The mythological son of Aeson and leader of the Argonauts. The witch Medea helped her lover Jason find the golden fleece by magic (*OCD*). Maier, *Atalanta,* commentaries to emblems XXV and XLIX; and idem, *Septimana philosophica,* emblem 4, present the legend of Jason as a model for alchemists (see de Jong, 307–9).

[82] For Hermes Trismegistus, see note 43. For Morienus, see note 65. Artephius, whose treatise was translated into English and published with Flamel's in 1624, is a mysterious figure whom Gildemeister (*Zeitschrift der morgenlandischen Gesellschaft* (1876), 538) thought to have been a twelfth-century Arabic philosopher. Legend has it that, by virtue of the elixir, he lived a thousand and twenty-five years (Ferguson, I: 51).

[83] *Juno.* The mythical wife of Jupiter who ruled aspects of women's lives. She is often associated with the moon (*OCD*).

[84] *serpents . . . the sage and wise man must strangle in his cradle.* A reference to Hera's attempt to murder Zeus's illegitimate child Heracles (Hercules) by sending serpents to attack him in his cradle. These he strangled, another allusion to overcoming the violent alchemical *putrefactio* (*OCD*).

[85] *Corassene bitch . . . Armenian dogge.* De Jong (285–9) traces the Corassene bitch and the Armenian dog as metaphors for the alchemical opposites mercury and sulphur to Rhasis, *Consilium conjugii,* cited in Petrus Bonus, *Margarita Pretiosa Novella Anno 1330* (*Theatrum chemicum,* 5:633). The two battling dogs are illustrated in Maier, *Atalanta,* emblem XLVII and Lambsprinck, *De lapide philosophico,* Emblem V.

[86] *permanent water.* "Enduring water, is that which is made by the philosophical solution out of two perfect metallic bodies . . . a sharp, penetrating water, which dissolves bodies" (Rulandus, 34–45).

[87] *Abridgement of Phylosophy.* Flamel's *Summarium philosophicum* (See note 59).

[88] *Rhasis.* See note 72.

[89] *Avicen.* Avicenna is the Latin name of Abu 'Ali al-Husain ibn 'Abdallah ibn Sina, a physician who was born in Bokhara c. 980 and died in 1037. The chemical writings attributed to Avicenna (i.e., *De congelatione et Conglutinatione Lapidum, Declaratio Lapidis Physici Avicennae Filio suo Aboali, Tractatulus de tinctura metallorum* and others) were probably not written by him, but by someone at least a century later (Ferguson, I:60).

[90] *Reynes.* Or Reins: kidneys (*OED*).

[91] *Democritus.* Democritus of Abdera (c. 460–370 B.C.), called the "laughing philosopher," was one of the foremost proponents of atomism.

[92] *triacle.* "A medicinal compound, originally a kind of salve, composed of many ingredients, formerly in repute as an antidote to venomous bites, poisons generally, and malignant diseases" (*OED*). "antidote," 1692; "*theriaque*," 1659.

[93] *Babylonian Dragon.* Revelations 12:3; 13:3, 11; 16:13.

[94] "*Flavastres*," 1659. Golden yellow.

[95] *Pontique.* Mercury.

[96] *Ship of Theseus* . . . The adventures of Theseus are another metaphor for the dangers of *putrefactio.*

[97] *head of the Crow.* Symbolizes the cleansing and whitening of *putrefactio*, also synonymous with the alchemical hermaphrodite.

[98] *coator permanent.* "a permanent or fixed water," 1692.

[99] *Erituration.* "*Trituration*," 1659. To grind or masticate.

[100] *Xir.* "The Matter of the Great Work when it has reached the Black Colour, and is thus in the state of Putrefaction" (Rulandus, 463).

[101] *Poulets head.* "top or head with the Vapour or fume," 1692. The top of the vessel.

[102] *naturall moysture.* "*humeur*," 1659.

[103] *Leton.* Practically speaking, "Lato," "Laton" or "Leton" is a yellow alloy such as brass (Kelly, 172). Rulandus (205–6) describes "Laton" as "An impure Red Body . . . Laton is Copper tinctured with Gold by means of the Stone Calamine . . ." Allegorically, Laton is the

prima materia lapidis in a state of baseness. Whitening, or purifying, the laton is the subject illustrated in Maier, *Atalanta*, emblem XI. De Jong (113–119) traces the source of the image to Morienus, *De transmutatione metallorum* (*Artis auriferae*, II:43–4).

[104] *Cadmus.* Cadmus, in Greek mythology, killed the serpent that guarded the Castalian Spring and sowed its teeth, which grew into armed men (*OCD*). The alchemical image of Cadmus piercing the serpent against a hollow tree appears in a seventeenth-century manuscript entitled *Speculum veritatis*, Biblioteca Apostolica Vaticana, Cod. lat. 7286, fol. 9.

[105] *rowles.* "mottos," 1692.

[106] *Painter.* "*sculpteur,*" 1659.

[107] *Chaos.* "The Unformed Matter, and the Confused First State of all Things" (Rulandus, 98).

[108] *kinsman.* "*parent,*" 1659.

[109] *Hermaphrodite of the Ancients.* "A name which the Philosophers have given to the purified matter of their Stone after Conjunction. It is properly their Mercury, which they call male and female" (Rulandus, 336).

The hermaphrodite appears in both manuscript and printed alchemical sources as a metaphor for the merging of the opposites into one incorruptible unified substance. Examples of manuscript images are: *Book of the Holy Trinity*, Nuremberg, Germanisches Nationalmuseum, MS 80061, fols. 99 and 100; *Aurora consurgens*, Zurich, Zentralbibliothek, MS Rh. 172, title page verso; *Rosarium philosophorum*, St. Gallen, Stadtbibliothek Vadiana, MS 394a, fol. 97; Printed sources are: *Rosarium philosophorum*, in *Artis auriferae*, II, 359; Mylius, *Philosophia reformata*, 354, fig. 5; and idem, *Opus medico-chemicum*, seals of "Aloysius Marlianus Philosoph," "Garsia Cardinalis Philosophus," "Rocherius Bacon Anglus Philos," "Hortulanus Philosophus Chymicus" and "Turba Philosophorum ac sapientum"; Jamsthaler, 75; Maier, *Symbola aureae mensae*; Maier, *Atalanta*, Emblem XXXIII; Mylius, *Philosophia reformata*, emblems 5, 7, 8, 10, 13, 14, and 17.

[110] *natures.* "*elements,*" 1659.

[111] *wombe of the Vessell.* Womb imagery is illustrated in Mylius, *Opus medico-chymicum*, seal of "Melchior Cibinensis Ungar, Philof," which is accompanied by the motto *Philosophic Philosophorum lapis ut infans, lacte nutriendus est virginali.*

[112] *quintessence.* "Nature, Potencies, Virtue, Tincture, Life, Spirit, the Medicine itself, and the quality of substances separated by Art from the body" (Rulandus, 272).

[113] *nourishment.* "The stone, like the infant, must be nourished" (*La Tourbe*, 29).

[114] *vegetable soule.* "Saturn is described as vegetal, not as a vegetable, and it is so called . . . because it possesses a vegetative soul, which cooks, digests, and conducts it to perfection" (Rulandus, 435).

[115] *Azoth.* "Azoth is quicksilver, drawn out of any metallic body, and properly the corporeal Mercury . . . When the Laton is whitened, it is called Azoth. Therefore men say Azoth whitens the Laton, then the Laton again whitens other things" (Rulandus, 66). Also the term for the universal medicine, the philosophers' stone, or the successful product of the alchemical work.

[116] *deluge.* "By this term the Philosophers understood the Distillation of their Matter, which, after it has ascended in the form of vapours to the summit of the vessel, returns upon the earth like a rain which completely inundates it" (Rulandus, 355).

[117] *God Apollo.* Greek sun god who ruled music, archery, prophecy, medicine and the care of flocks and herds. Apollo's earliest adventure was the killing of Python, a formidable dragon which guarded Delphi (*OCD*). The Python-Apollo simile is here a metaphor for the alchemist's vanquishing of *putrefactio.* Maier, *Atalanta*, Emblem XXV, illustrates Apollo hunting Python in the same visual space as sol and luna conquering the alchemical dragon.

[118] *moity.* Half.

[119] *Theseus.* The labors of the alchemist to resolve the *putrefactio* stage are here compared to the exploits of Theseus, who killed the minotaur and defeated the Amazons.

[120] *Nummus.* "Lead, Black Lead" (Rulandus, 240).

[121] *Ethelia.* "Ethel is the Black, also the Fire. The *Turba* says: Ethelia is the Burnt Body, parched and dry, red and white, fire and sieve, or riddle, which holds together the water of the Mercury" (Rulandus, 137).

[122] *Arena.* Sand (Rulandus, 37).

[123] *Boritis.* "The White Stone after the Black Slate. It reduces earth to water" (Rulandus, 75).

[124] *Corsufle.* "The sulphur of the Philosophers fixed at the Red Stage" (Rulandus, 352).

[125] *Cambar.* "The Matter of the Sages arrived at the white stage" (Rulandus, 345).

[126] *Duenech.* "Antimony" (Rulandus, 129). Dueneg, duenez, duenum or duenic, is the Arabic word for vitriol, a solvent for metals.

[127] *Randeric.* "*Bauderick,*" 1659. "Matter of the Work, or Rebis, before it has attained the white stage" (Rulandus, 416).

[128] *Kukul.* "Kuhul . . . is the Lead of the Philosophers" (Rulandus, 187).

[129] *Thabricis.* "*Thabritris,*" 1659; also Thabritius, Gabricus or Kybric, all synonyms for sulphur.

[130] *Ebisemech.* "The Matter of the Hermetic Chemists during its period of putrefaction" (Rulandus, 358).

[131] *Ixir.* See note 100.

[132] *Phylosophers Ladder.* The ladder symbolized the successive steps of the alchemical work. Perhaps the earliest use of the metaphor is Zosimos' description of the ascent and descent of the fifteen steps of light and darkness (Berthelot, III, i, 2). The ladder is illustrated in "Emblematical Figures of the Philosphers' Stone," London, British Museum, MS Add. 1316; *Adamas colluctancium aquiliarum*, Biblioteca Apostolica Vaticana, MS Pal. Lat. 412, fol. 77 verso; Paris, Bibliothèque Nationale, MS Fr. 14765, fol. 214; Altus, *Mutus liber*, 15; and Mylius, *Opus medicochymicum*, seal of "Johannes Mehum Philosophus."

[133] *haire.* "*capillaire,*" 1659.

[134] *cierine.* "*subcitrine,*" 1659.

[135] *Lynceus.* The Greek mythological figure whose sight was preternaturally sharp and who looked out for the approach of the twins Castor and Pollux, sons of Zeus (*OCD*).

[136] *colour of perfection.* Red.

[137] *Allegory of Aristeus.* Jung (*Mysterium Conjunctionis*, 513) notes the Plinian legend that Aristeas saw his soul fly out of his mouth in the

shape of a raven. Ravens and black birds are common alchemical symbols for putrefaction.

[138] *troubles of the Sea.* Dark, heated, dangerous waters are further synonyms for *putrefactio.*

[139] *our King.* The image of the lapis-King appears in many illustrated alchemical treatises, notably throughout Lambsprinck, *De lapide philosophica libellus*; Maier, *Atalanta*; Michelspacher, *Cabala*; Mylius, *Opus medico-chymicum* and *Rosarium philosophorum.* See also the 15th-century manuscript of Johannes Andreae, London, British Library, MS. Sloane 2560, fol. 15.

[140] *Albification.* Whitening, cleansing.

[141] *Turba Phylosophorum.* An anonymous collection of quotations from ancient authors. Some believe it to be Arabic in origin, and others date it from the eleventh century. Thorndike (3:41) notes its introduction into alchemical treatises beginning in the thirteenth century, and this theory is born out by the astounding number of manuscripts which quote the *Turba* from this time onward. The treatise existed in manuscript form until 1572, when it was printed.

[142] *Seethe.* "decoct," 1692.

[143] *head of the Crow.* See note 97.

[144] *Lambspringk.* Little is known about the alchemical poet Lambsprinck, though some think he was a Benedictine attached to the Abbey of Lammspring, near Hildesheim. His treatise *De lapide philosophico libellus* probably originated in the early fifteenth century, and was eventually printed in Latin and German in the seventeenth century. The emblems that illustrate the treatise are similar to those described by Flamel (Ferguson, II:5).

[145] *Periphrasis.* A roundabout way of speaking; circumlocution (*OED*).

[146] *Serpent Hydra.* Reference to the second labor of Heracles, when he slew the Hydra of Lerna, whose heads regrew after being cut off (*OCD*).

[147] *white stone.* "When the water covers the earth, then the white is over the black . . . Therefore it is called the White . . . Lead, Exebmich, white Bismuth, Martech, white Ore, white Stone" (Rulandus, 204).

[148] *imbibe.* "Ablution, when a liquid joined to a body is made light, and, finding no exit retreats into the body, and washes it so long with frequent illustrations, until it is wholly coagulate therewith, and is unable to rise further, but the whole remains fixed" (Rulandus, 182).

[149] *Salamander.* The firey salamander, who was thought capable of surviving in all four elements, is illustrated in Maier, *Atalanta*, Emblem XXIX; Mylius, *Philosophia reformata*, emblem 4 and Lambsprinck, *De lapide philosophico*, emblem 10.

[150] *Virgins milke.* "*Lac Virginis* is Mercurial Water, the Dragon's Tail; it washes and coagulates without any manual labour; it is the Mercury of the Philosophers, Lunar and Solar Sap, out of catholic earth and water" (Rulandus, 188).

[151] *leprous.* A synonym for *putrefactio.* (See Pernety, *Dictionnaire*, 245).

[152] *Naaman the Syrian.* The Biblical Naaman the Syrian was a commander of the Syrian army who contracted leprosy. On the advice of the prophet Elisha, he bathed seven times in the River Jordan, was cured and converted (Kings 2:5).

[153] *Corsufle.* See note 124.

[154] *teine.* Tinge.

[155] *Senior.* The treatises of Senior (Mohamnmed ibn Umail al-Tamini, ca. 900–60) were translated into Latin in the early twelfth century.

[156] *Infants belly.* The infant is the alchemical Mercurius, whose self-generating persona is described by Petrus Bonus as "father before being son," who ". . . creates my mother and my father" after being born from them (Bonus, 235).

[157] *Hercules hath clensed the stable.* . . . Reference to the sixth labor of Hercules when he cleaned the stables of Augeas (*OCD*). Here cleansing, or albification, of the materials in *putrefactio* is compared with Hercules' cleansing of the filth of the stables.

[158] *Amalthaea.* Amalthaea, a nurse of Zeus, was a she-goat whose horns flowed with nectar and ambrosia. When one of them broke off, it was filled with fruits and given to Zeus. The legend is the origin of the *cornu copiae* (*OCD*).

[159] *Booke of the seven Egyptian seales.* No ancient text by this name is known. However, the passage probably refers to the *Book of Abraham* itself, whose seven emblems are called "hieroglyphics" in the text. Michael Maier, in 1614, published *Arcana arcanissima* (London) which is believed to be the first hermetic interpretation of Egyptian myths.

[160] *vermillion.* "*de Laque*," 1659. Like laquer.

[161] *Virgin's milke of the Sunne.* See note 150.

[162] *vert.* Green.

[163] *tree.* "*germe*," 1659.

[164] *white Stone.* See note 147.

[165] *white Elixir.* An extract or tincture capable of transmuting base metals into silver.

[166] *coppell.* Cupel. A small shallow porous cup, usually made of bone-ash, used in assaying gold or silver with lead (*OED*).

[167] *seven times refined.* Malachi 3:2–3, ". . . for he is like a refiner's fire . . . And he shall sit as a refiner and purifier of silver: and he shall purify the sons of Levi, and purge them as gold and silver . . . "

[168] *haultboy. Hautbois,* or shawm, the ancestor of our modern oboe. The object illustrated here, however, is not a shawm, but a bagpipe. The confusion in labeling the instrument arises from the fact that sound is produced in both shawms and bagpipes by blowing air through a double-reed pipe. Shawms and bagpipes are illustrated together in Michael Praetorius, *Theatrum instrumentorum* (1620) and co-existed in a consort of wind instruments in Louis XIV's *Grande écurie du roi* (See Sibyl Marcuse, *A Survey of Musical Instruments* (New York, 1975), 686, 696).

The choice of musical instruments also recalls the duality of body and spirit inherent in alchemical iconography. The lute, a soft stringed instrument, would represent the soul, and the bagpipe, a loud, raucous-sounding instrument rarely found in the hands of angels, would represent the earthly, bodily realm.

[169] *Peacocks taile.* Alchemy recognized the peacock's tail, or *cauda pavonis,* as the multiplicity of colors appearing just before the "whitening" that preceded transmutation. The *cauda pavonis* announces the end of the work, like its symbolic synonym Iris, messenger of the gods (see note 31). The peacock is described in George Ripley's "Cantilena,"

Opera omnia chemica (Cassel, 1649), 421 ff., verse 17, trans. F. Sherwood Taylor, *Ambix* II (December, 1946). The peacock's tail is illustrated in manuscript versions of Trismosin's *Splendor solis* and in many printed books. See: Boschius, *Symbolographia*, Symbol LXXXIV; Khunrath, *Amphitheatrum sapientiae*, Fig. 8; Valentine, *Duodecim clavibus*, 9th key; and Mylius *Opus medicum chemicum*, seal of "Eduardus Kellae, Philosoph, Dubius."

[170] *Poppey of the Hermitage.* "corn poppy," 1692.

[171] *vermillion. "laque,"* 1659.

[172] *man with a key.* St. Peter, who traditionally holds the keys to heaven, is compared to the successful alchemist holding the "key" to transmutation. Illustrations of alchemical "keys" appear in Biblioteca Apostolica Vaticana, MS Pal. Lat. 412, fol. 44 verso; and Leiden, Bibliotheek der Rijksuniversiteit, Cod. Voss. Chym. F. 29, fol. 99.

[173] This description of the man and lion is realized and illustrated as the seal of "Nicolaus Flamellus Gallus" in Mylius, *Opus medicochymicum.*

[174] *red . . . flying Lyon.* Probably a symbol for volatile red sulphur (See Mylius, *Philosophia reformata,* 209).

[175] *Granadored. "granade,"* 1659. "cinabar Red," 1692.

[176] *Shee.* "to wit, the Stone, Elixir, or Tincture," 1692.

[177] *pleasant colour.* "Papaveran red," 1692.

[178] *poppy of the Rocke. "Sylvestre,"* 1659.

[179] *Tyrian . . . colour.* A reference to the purple or crimson dye made at Tyre in ancient times from certain mollusks (*OED*).

Artephius His Secret Booke

[1] On Pontanus' life, see intro., pp. xl–xli. As claimed in the title, the *Epistle* appears in *Theatrum chemicum* (Strasbourg, 1613), 3:734.

[2] *Quintessence.* The most essential part of any substance, a highly refined essence or extract (Kelly, 179).

[3] *Roger Bacon.* A learned Franciscan who contributed to all branches of science. Born circa 1214, Bacon studied at Oxford before being sent to Paris in 1257. He authored several significant alchemical treatises. The *"Libro de mirabilibus naturae ae operibus,* or "Book of the wonderfull workes of nature" mentioned here was first published in 1542, then again in 1594. A revised and corrected edition appeared at Hamburg in 1618 and an English translation was published in 1597 (Ferguson, I:64–6). See Linden, *Mirror of Alchimy*, intro.

[4] *Paracelsus.* Born Philippus Theophrastus Bombast von Hohenheim in the year 1493, Paracelsus was famous for declaring that the sole purpose of alchemy should be to make medicines. He applied both chemical and herbal cures with great success, though his difficult personality and arrogant stand against the ancients caused him to be banished from Basel.

[5] *this Kingdome.* "Royaume de France," 1659.

[6] *Antimony.* Antimony trisulphide, also known as "stibnite" (Kelly, 177).

[7] *Saturne.* Lead (Rulandus, 287).

[8] *Saturnine Antimonie.* Antimonium . . . is Dross of Lead, and has the virtues of burnt Lead, being of similar substance. It is cold, dry, and

astringent (Rulandus, 32).

[9] *Argent vive.* Mercury (Rulandus, 46).

[10] *Gold.* "*Soleil*," 1659. Throughout Artephius' treatise, the words "gold" and "sun" are used interchangeably.

[11] *glasse.* "*miroir*," 1659. "looking glass," 1692.

[12] *Vinegre Antimoniall.* Antimonious acid. The making of "vinegar of Antimony" and its properties are described in Basil Valentine, *Triumphant Chariot of Antimony* (London, 1678).

[13] *Sal Ammoniack.* corr: Sal armoniack, normally ammonium chloride (Kelly, 179).

[14] *foure fingers.* "four inches," 1692.

[15] *in manner of a thinne skinne.* "much like a Scum," 1692.

[16] *make the water vapour away by the fire.* Evaporate.

[17] *Magnesia.* "Artificial magnesia is melted tin when mercury has been injected into it, and the two have been mingled together until they form a brittle substance, and a white mass" (Rulandus, 216).

[18] *Argent vive.* See note 9 above.

[19] *Leton.* See note 103, Flamel.

[20] *Sunne and . . . Moone.* The alchemical opposites.

[21] *putrifieth as a graine of Corne.* The comparison between chemistry and agriculture originates with (Pseudo) Aristotle, *Tractitatulus Aristotelis de Practica Lapidis Philosophici*, in *Artis auriferae*, I:398, "Seminate aurum vestrum in terram albam foliatam" (Sow your gold in the white, foliated earth). Rulandus (423) says "To Sow is to Cook, i.e. to Continue the Regimen of the Fire." Valentinus, *Practica cum duodecim clavibus*, Fig. 8, and Maier, *Atalanta*, Emblem VI, illustrate the "sowing" of gold like seeds in the earth, where it will putrefy and multiply (See de Jong, 81–7).

[22] *silver medicinall.* "a medicinal white gold," 1692.

[23] *tincture and great fusion.* "melting or dissolving," 1692.

[24] *disolve.* "liquifying or melting," 1692.

[25] *mettall.* "Leaves, Filings, or Calx of any mettall," 1691.

[26] *oyle.* Perhaps concentrated sulfuric acid, called "oyl of sulphur" and "oyl of vitriol" (Kelly, 179).

[27] *Balneum Mariae.* See note 55, Flamel.

[28] *amended.* "to be exalted," 1692.

[29] *Bodies.* Metals.

[30] *Sal Albroe.* "a fusible Salt," 1692.

[31] *nature.* "fixedness," 1692.

[32] *bloudy.* "blood-color-making," 1692.

[33] *water of May-dew. Aqua vitae,* or mercury (Rulandus, 36).

[34] *Azot.* For Azoth, see note 115, Flamel.

[35] *wombe.* For alchemical womb imagery, see note 111, Flamel.

[36] *Fountaine in which the King and Queene wash.* . . . For the imagery of the King and Queen bathing, see note 20, Flamel.

[37] *sublimation.* Vaporization without passing through a liquid phase (Kelly, 180).

[38] *middle.* Neutral.

[39] *Corsufle.* See note 124, Flamel.

[40] *Cambar.* "The matter of the Sages arrived at the white stage" (Rulandus, 345).

[41] *Ethelia.* See note 121, Flamel.

[42] *Zandarach.* "Zandarith," 1692; "A Substance which participates in equal proportion of Body and Spirit, that is, of the Fixed and the Volatile" (Rulandus, 465).

[43] *Duenech.* See Flamel, note 126.

[44] *water permanent.* "Enduring Water, is that which is made by the philosophical solution out of two perfect metallic bodies. It is Sol and Luna dissolved in water, and likewise united" (Rulandus, 34–5).

[45] *tincture.* To penetrate and permeate with color (Rulandus, 318).

[46] *uncleannesse of the dead.* A synonym for putrefaction.

[47] *fœces.* Filth and impurities.

[48] *golden Fleece.* The philosophers' stone. For reference to Jason in alchemical terms, see notes 41 and 81, Flamel.

[49] *Menstrous and corrupt place of his original.* "Menstruum is that from which all metals are derived" (Rulandus, 228).

[50] *Hermes his Bird.* The volatile spirit *mercurius.*

[51] *red servant.* The *Turba philosophorum* says, "Join that male who is the son of the red slave with his fragrant wife . . . How exceeding precious is the nature of that red slave, without which the regimen cannot endure." The substance called the "servant" was probably common quicksilver which became exalted by alchemy (*La Tourbe*, 96–7).

[52] *Diers.* "Colours in Dying Cloth," 1692.

[53] *Adsar.* A legendary Alexandrian Christian alchemist who was the teacher of Morienes who was, in turn, the teacher of Khalid (Thorndike, 2:214, 216).

[54] *for governing it in its bath.* "for being digested in Balneo (Mariae)," 1692.

[55] *the vulture flying without wings . . . I tell truth, and lie not.* The wingless vulture is the symbol of fixed philosophical mercury. The multicolored vulture crying from the mountain top originates with Hermes Trismegistus, *Tractatus vere aureus, de lapidis physici secreto* (*Theatrum chemicum*, 4:618) ". . . quod vultur super montem existens, clamat voce magna: Egpo sum albus nigri et rubeus albi et citrinus rubei et certe veridicus sum" (. . . the vulture living on the mountain, calls in a loud voice: I am the white of the black, and the red of the white and the yellow of the red . . .). The epigram accompanying Emblem XLIII of Maier's *Atalanta* also says, "The vulture is standing on the top of a high mountain, incessantly calling: It is said that I am white and black, Yellow and red and I do not lie at all" (See de Jong, 268–72). Here, the colors of alchemy are four: black for putrefaction, white for albification and yellow (*citrinitas*) which precedes the final red of the *lapis.*

[56] *conjunction.* Joining together of the alchemical opposites.

[57] *assation.* "Assatio . . . to make into a hard and dry Ash" (Rulandus, 51).

[58] *subtiliation.* "Subtilatio, Subtilation . . . A Dissolving or separating of the subtle parts from the gross" (Rulandus, 303).

[59] *Ariadnes thread.* Ariadne, in mythology, is the daughter of Minos and Pasiphae. When Theseus came to Crete, she fell in love with him and gave him a clue of thread by which he found his way out of the Labyrinth after killing the Minotaur (*OCD*).

[60] *Hesperides.* See note 80, Flamel.

[61] *Balneum Mariae.* See note 55, Flamel.

[62] *worke of Women.* Rulandus (462) defines "Work of the Woman" as "The Great Work; a term used, as we have seen, on account of the facility with which the Stone may be composed by those who are instructed in the proper method of operation." Maier, *Atalanta*, emblem XXII, illustrates a woman performing the alchemical work accompanied by the words "Plumbo habito candido fac opus mulierum, hoc est, COQUE" (Once you have the white lead, do women's work, that is, COOK). The text explains that the alchemist should treat his ingredients in the same way as a woman who cooks fish: by softening what is hard, dissolving and evaporating by boiling, and taking care that the substance be not burnt. De Jong (178–9) traces the origin of the allegory to the *Turba philosophorum* (*Artis auriferae*, I:10), ". . . Nunc autem plumbi albi dispositionem monstravi, quo noto nihil aliud sequitur, quam Opus mulierum et Ludus Puerorum" (Now, however, I have shown the systematical arrangement of the white lead, and when that is known, nothing else follows but women's work and child's play.")

[63] *Play of Children.* Related to the "work of women" in that it shares the same origin in the *Turba philosophorum* (*Ibid.*). The *ludus puerorum* allegorically describes the playful fermentation and bubbling of the ingredients requiring little attention from the practitioner. Trismosin (*La Toyson d'or*, MS Fr. 12297, fol. 54, Paris, Bibliothèque Nationale) describes it as ". . . compared to a children's game, or the pleasure and high spirits of children doing frivolous things." "Child's play" is illustrated in a manuscript version of Trismosin's *Splendor solis*, London, British Museum, MS. Harley 3469 and in printed editions of the work.

[64] *place where the King and Queene bathe.* . . . See note 20, Flamel.

[65] *flowers.* See note 66, Flamel.

[66] *calcines.* "Calcinatio . . . Combustion, which takes place in a strong heat . . . It is of two kinds: That of Corrosion and of Ignition"

(Rulandus, 87).

[67] *dominion of the woman.* Reference to the *mensis philosophicus,* the dark, moist period of putrefaction that imitates the movements and attributes of the Moon (Rulandus, 228).

[68] *the thing which hath the head red and white, etc.* A reference to the three major color changes in alchemy: black (or blue) = putrefaction; white = cleansing; and red, the color of the *lapis.* See note 60, Flamel.

[69] *house of glasse.* A sealed glass flask. Maier, *Atalanta,* Emblem IX, illustrates the *Lapis* as an old man locked in a furnace-shaped house. The accompanying motto says "Lock the tree with the old man in a bedewed house." De Jong (102), traces the source of the motto to the *Turba philosophorum* (*Artis auriferae,* I:57) "Accipe illam arborum albam, aedifica ei domum rotundam, tenebrosam, et rore circundatam, et impone ei magnae aetatis hominemcentrum annorum, et claude domum, ne ventus aut pulvis ad eos perveniat:" (Take that white tree surrounded by dew, build around it a round, dark house, put in it a man stricken in years, a hundred years old, and lock the house, so that no wind or dust may penetrate to them).

[70] *Sonne of the Virgin . . . lifted up towards the heavens . . .* A comparison of the alchemical *lapis* to Christ, who, like the *lapis,* was punished, died and resurrected to immortality.

[71] *Argent vive.* See note 9 above.

[72] *fœces.* Impure dross.

[73] *stormy Sea.* See note 41, Flamel.

[74] *fat of the soule.* See note 52, Flamel.

[75] *Toade going upon the Earth.* The toad and the vulture are synonyms for the fixed and volatile components of alchemy. An image of a flying vulture chained to an earth-bound toad appears in Maier, *Symbola aureae mensae* illustrating Avicenna's dictum that the volatization of the fixed and the fixation of the volatile, constitute the entire Work.

[76] *as the rennet of a Lambe coagulateth milke into Cheese.* The making of cheese in the stomach of a lamb refers not only to the making of the *lapis,* but to the resemblance of the conception and birth of the *lapis* to human generation, which was believed to involve the coagulation of menstrum and semen in the same manner as cheese forms into solid curds.

[77] *clense the vulture.* The impure vulture is a symbol for Philosophical Mercury in putrefaction. See Maier, *Atalanta*, commentary on Emblem XLIII.

[78] All are metaphors for *putrefactio.*

[79] *stinking smell . . . of graves.* Perhaps the "smel of sepulchres" refers to the rotten-egg odor of sulphur in putrefaction.

[80] *Poulet of Hermogenes.* See note 60, Flamel.

Pontanus, Epistola

[1] For the life of John Pontanus, see intro., pp. xl–xli. Besides the *Epistle*, he wrote "Methodus componendi theriacam et praeparandi ambram factitiam," in Johann Wittichius' *Consilia observationes et epistolae medicae . . . Collecta* (1604) and "De prodigiosis episcopi spirensis jejuniis," in Lentulus' *Historia admiranda de prodigiosa Apolloniae Schreierae* (Bern, 1604) (Ferguson, II: 212–13).

[2] *Giber.* corr., Geber. Ferguson (I:302) suggests that Geber was the Arabian alchemist Jabir ibn Hayyan. However, Berthelot noted that the Latin works of Geber have nothing to do with the Arabic works of Jabir, but are considerably later. Whatever the case, the name "Geber" first appears in alchemical manuscripts of the fourteenth century. Geber's chemical writings, which emphasize practical laboratory procedure and apparatus, were first published c. 1475.

[3] *Archelaus.* One of the Greek philosophers included in the *Turba philosophorum* (See note 141, Flamel).

[4] *that which is above, and that which is below.* A partial quotation of the second precept of Hermes Trismegistus' *Emerald Tablet* (*Tabula smaragdina*, reproduced and translated in J. Read, *Prelude to Chemistry*, 54): "What is below is like that which is above, and that which is above is like that which is below, to accomplish the miracles of one thing."

[5] *Arnold.* Arnauld of Villanova was born c. 1234 or 1250 and died in 1309. He was a physician, alchemist and astrologer whose treatises, which were first printed in 1504, are medically oriented and speak against magic and sorcery (Thorndike, 2:cap. LXVIII).

Index